**Models of
Industrial
Structure**

Models of Industrial Structure

Lars Engwall
Stockholm University

Lexington Books
D.C. Heath and Company
Lexington, Massachusetts
Toronto   London

Library of Congress Cataloging in Publication Data

Engwall, Lars.
  Models of industrial structure.

  Bibliography: p.
    1. Industrial organization.   2. Industries, size of.
  3. Stochastic processes.   I. Title.
  HD2731.E66   658.1   72–7021
  ISBN 0–669–85167–1

Published simultaneously in Canada.

Printed in the United States of America.

International Standard Book Number: 0–669–85167–1

Library of Congress Catalog Card Number: 72–7021

Contents

Part 4: Conclusions

List of Figures

List of Tables

Preface

This book is based on my dissertation *Size Distributions of Firms.* Those familiar with my thesis will find the major changes in chapters 2, 3 and 9. They will also find that a chapter on simulation—chapter 7—has been added. Moreover, new empirical applications are presented in chapters 6, 8 and 9. The addition to chapter 6 concerns industrial structure in socialist countries, whereas chapters 8 and 9 include results from additional computer applications. In chapter 9 these applications also contain an example showing the use of subjective probability estimates, a technique also discussed in the chapter. Those who have come across my published articles will recognize some of the chapters. Thus earlier versions of chapters 4, 5 and 8 have been published in *The Swedish Journal of Economics* (1968), *Metroeconomica* (1970) and *The Canadian Journal of Economics* (1970), respectively. Permission to use these articles is hereby acknowledged with gratitude.

The main part of this study was carried out at the Department of Business Administration, Stockholm University, Sweden. There I am much indebted to Professor Bertil Näslund, who first interested me in the field of research and who very willingly helped me with many problems during the whole course of my study. Comments from my colleagues at the department constituted another valuable source of improvements on preliminary drafts of the chapters.

During a year's leave from Stockholm University, I was able to work on this project at The Graduate School of Industrial Administration, Carnegie-Mellon University, Pittsburgh, Pennsylvania, and at the University of California at Berkeley, Berkeley, California. These institutions provided a very stimulating milieu for my research, and I highly appreciated contacts with various faculty members.

Many persons have helped me in various phases of this study. I owe all these unnamed persons many thanks.

A grant from The Bank of Sweden Tercentenary Fund has been the major financial source for this study. Grants from Magnus Bergvalls Stiftelse, Carl-Bertel Nathorsts Stiftelse, Stiftelsen Lars Hiertas Minne, Sparbankernas Forskningsstiftelse, Sverige-Amerika Stiftelsen and Söderströms Fond made my stay in the United States possible. I am much indebted to all these institutions for their support.

**Models of
Industrial
Structure**

Introductory Summary

Chapter 1: Stochastic Models and Changes in Industrial Structure. The chapter first sketches the rapid changes in economic structure after World War II. The transfer of resources from agriculture and forestry to industry and service activities is mentioned. Then changes in industrial structure are discussed.

Following these historical facts, the concept of concentration as well as the means of analyzing concentration are introduced. In this context stochastic theories—the approach used throughout the book—are presented. Finally, the chapter gives a brief outline of the book.

Chapter 2: Some Basic Concepts. This chapter deals with a number of concepts essential to the study: firm, size, and concentration. First of all, however, the data sources of the study are described.

The data used comprise samples of (1) large firms in five areas, (2) establishments manufacturing cars and shoes in Sweden, and (3) enterprises in four socialist countries.

As for the firm concept, several different alternatives—plant, establishment, enterprise, etc.—are mentioned and discussed in relation to the samples.

Regarding size, a distinction is made between measures of (1) input, (2) output, and (3) input and output. It is concluded that output measures are preferable in most cases, if they are readily available. In order to check the relevance of using input measures as a proxy for output measures, rank correlation coefficients are computed. They indicate a high degree of correlation between measures of output and input.

In reviewing measures of concentration, reference is made to earlier works from the present century. This review starts by summarizing measures based on mean and variance and proceeds to measures using market shares. Measures belonging to the latter group are the concentration ratio, the Lorenz curve, the Gini coefficient, Herfindahl's index, and entropy.

In an empirical section the characteristics of the different measures are compared.

Chapter 3: Analytical Models in Review. In this chapter a distinction is made between (1) models using discrete size, and (2) models using continuous size. In a model of the first kind, changes in size are represented by jumps between classes. In the continuous models, on the other hand, the changes in size are percentage changes in present sizes.

As for discrete models, the underlying assumptions as well as some basic properties of a finite Markov chain are discussed.

Regarding the continuous models, interest is focused on their assumptions. Particular attention is given to the law of proportionate effect, often associated with one of the continuous models. It implies that all firms in a population of firms have the same probability of growing a certain percentage regardless of their size.

Concerning both kinds of models, reference is made to earlier works as well as to subsequent chapters.

Chapter 4: A Discrete Model Assuming the Law of Proportionate Effect. The model introduced in this chapter is supposed to be valid for the largest firms within an area. It is based on two assumptions, i.e., (1) that the law of proportionate effect is valid above a certain size, and (2) that new firms are born at a relatively constant rate in the lowest size class. Under these assumptions, it can be shown that the process is moving toward a steady state. The distribution of firms in this state can be approximated by the Yule distribution.

The first two sections of the chapter refer to the model and its assumptions. Then a description of different test procedures follows. These methods are applied in the empirical section, where the samples of large firms are investigated. This analysis reveals no reason to reject the model for the samples mentioned. However, validity seems to be limited to firms of extreme size; the number of these firms does not exceed 200 in any of the five areas investigated.

From one of the parameters of the distribution it is possible to derive the ratio between net growth attributable to firms above the minimum size and the net growth of all firms. This ratio is the largest for the sample outside the United States and for Europe (40–50 percent), and smallest for the samples from the United States and Sweden (about 30 percent).

Chapter 5: The Transition Process. This chapter deals with transition matrices as a tool for describing changes in industrial structure. After an introduction, methods of estimating transition probabilities are discussed. Two methods are mentioned, i.e., (1) the maximum-likelihood method and (2) the least-square method. In subsequent sections steady-state distributions and measures of mobility are considered.

In an empirical section the samples of large firms are studied. Here it can be seen that the probability is greatest that a firm will remain within a given class, followed by the probability of transition to an adjacent class. Other probabilities are mostly estimated at zero.

During the investigated period 1956–65, movements toward a steady state can be observed. However, some differences between actual and theoretical distributions still exist.

The average time spent in the different classes varies around ten years. An index for mobility indicates that actual mobilities are about 10–20 percent of that in a prefectly mobile industry. The values of the index fairly well follow a ranking based on Gini's coefficient of concentration.

Chapter 6: Models Using Continuous Size. In this chapter two analytical models using continuous size are examined. The models produce a distribution which is frequently used in describing size distributions: the lognormal distribution.

After a review of earlier works, the models are described and their assumptions discussed. In a subsequent section different procedures for estimating parameters in a lognormal distribution are examined. Such estimates are a necessary prerequisite for the test procedures accounted for in a following section. Most of these procedures relate to other measures of concentration.

In an empirical section, the Swedish establishment distributions are tested for lognormality. The conclusion is that the lognormal distribution is a description which seems acceptable for industries showing moderate change in the number of firms. This conclusion is based on tests for lognormality as well as tests of the assumptions. Finally, establishment distributions in four socialist countries are tested for lognormality. No reasons to reject the lognormal description appeared.

Chapter 7: Simulation. This chapter serves as an introduction to the third part of the book, where two simulation models are presented. Following a general description of the technique—particularly computer simulation—some examples of simulation models are mentioned. Then the discussion is focused on the tradeoff between realism and simplicity in model building. In this context, problems involved in validation are given special attention. Finally, two earlier simulation models relating to size distributions of firms are summarized. They are models developed by Balderston & Hoggatt and Ijiri & Simon.

Chapter 8: A Simulation Model Using Discrete Size. This chapter contains a presentation of a model simulating changes in industrial structure. The model uses discrete size and is basically an extension of the model discussed in chapter 5.

The changes in size of existing firms are regulated by a transition matrix. Furthermore, another random process introduces new firms in the lowest class.

In the basic model all firms are influenced by the same mechanism, and the mechanisms are constant over time. However, the chapter also indicates possible modifications in the basic model. This means that certain aspects such as earlier growth rates, propensity to do research, the business cycle, and so on, can be included in the model.

After a description of estimation procedures, the validity of the model is checked. This is done by reproduction of historical conditions as well as by forecasting future conditions. These applications provided no reason to reject the model. As a result, the model was run in order to forecast the structure in 1980.

Chapter 9: A Simulation Model Using Continuous Size. A model using discrete size implies that only a rough measure of size is used and that some information on size is left out. In some cases more complete information on firm sizes in the distribution might be desirable. In these instances a simulation model using continuous size is appropriate. Such a model is presented in the present chapter. As in the preceding chapter, the description of a basic model is followed by descriptions of possible modifications.

The continuous model contains three basic components. They are mechanisms determining (1) the changes in size among firms that remain, (2) the entries of new firms, and (3) the exits of firms. This means that in comparison with the model presented in the preceding chapter one mechanism is added: the exit mechanism. In the discrete model this mechanism is included in the transition process; here it is a process in itself.

As for estimation procedures, two approaches are suggested. First, the estimation from historical data is discussed. Then the use of subjective probability estimates is examined. The last technique will be particularly useful in forecasting, as the present trend changes. In validating the model, reproduction of historical conditions as well as forecasting of future conditions are used. These applications did not give cause for rejection of the model, and consequently forecasting runs until 1980 were performed. Finally, an example of forecasting by means of subjective probability estimates is presented.

Chapter 10: Conclusions and Suggestions for Future Research.   This final chapter first summarizes the findings with respect to the methods and the empirical distributions. In other words, the contents of earlier chapters are integrated.

Then some suggestions for future research are given. Studies of entries and exits are given particular attention in this context. Moreover, some factors possibly related to changes in concentration are mentioned: economic policy, other economic distributions, education, and so forth.

Part I
The Topic and Related Concepts

1 Stochastic Models and Changes in Industrial Structure

After World War II, rapid changes occurred in the economic structure of most countries and in highly industrialized countries in particular. These changes imply transfer of resources from agriculture and forestry to industry and service activities. Also, a shift away from industries producing the necessities of life such as food, textiles, and shoes toward an increasing emphasis on engineering and chemical industries is in progress.

Many important factors, such as the expansion of international trade and the availability of more rapid transportation between countries, have contributed to the described transformation which has resulted in a great increase in the degree of international competition.

In many countries these changes have implied an increase in the large firms' share of the total amount of size units, a circumstance which is generally referred to as concentration. For instance, Scherer concludes that the share of value-added of the 100 largest manufacturing firms in the United States has increased from 23 percent in 1947 to 33 percent in 1966 (*Scherer*, 1970, p. 44). Similar conclusions have also been drawn for socialist countries (see *Woroniak*, 1970) as well as for a small country like Sweden (see *Carling*, 1968).

Tendencies toward an increasing degree of concentration have attracted vast attention on the part of economists. Consequently, many different measures have been suggested in order to measure industrial concentration over time.[1] This interest is not limited to the academic community, however. Politicians and businessmen have also called for increased knowledge of structural changes. Interest on the part of the former group is expressed in government sponsored research (see, e.g., *Carling*, 1968) and committee activities on the subject (see *Economic Concentration*, 1964–67). As for the latter group, the rapid changes outlined above point to an increased need to predict future events. This is reflected in a heavier stress on strategic planning in the firms as the speed of structural changes has increased. In this context the forecasting of future market structures is one important aspect (see *Ansoff*, 1965, p. 146) which can be expected to remain in the focus.

The interest sketched here is one reason to devote research to changes in industrial structure. Another is the controversies which have arisen concerning the interpretation of structural changes. As a matter of fact, there has been considerable disagreement in the politico-economic debate concerning the actual level of concentration and its changes over time. Consequently, it seems important to try to acquire more accurate measures of concentration as well as a better understanding of structural changes. The present study is directed toward different approaches to gain better knowledge on structural changes. In so doing, the emphasis will be put on stochastic models generating skew size distributions. They imply, generally speaking, that changes in size are regarded as a random process.

One main reason for choosing stochastic models is that they seem very promising for the topic selected. Another reason is that they are relatively simple.[2] This simplicity has two implications:

1.   The models are fairly easy to handle.
2.   Simple models are likely to become extreme, which means that they are easier to falsify. Consequently, the risk of accepting a false model is reduced (see *Popper*, 1959, p. 136 ff.).

The great part of earlier research on stochastic models has been performed during the last four decades. One of the earlier works is a study by *Gibrat* (1931), whose contribution has been very important for later research. However, it must also be mentioned that this study to some extent has been a drawback for later studies, as stochastic theories sometimes are evaluated on the plausibility of a principle introduced by *Gibrat* (1931, p. 62 ff.), namely, *the law of proportionate effect* (la loi de l'effet proportionnel). This principle implies that the probability of a certain change in size is independent of firm size. For instance, a firm having a sales volume of $10 million has, according to this law, the same probability to grow by $g\%$ as a firm having a sales volume of $100 million. It should be stressed here that this law is not an indispensable assumption for stochastic models. Consequently, some models assume the law to be valid, while others do not.

Among other earlier studies employing stochastic models of changes in industrial structure may be mentioned *Champernowne* (1937 and 1953), *Simon* (1955), *Hart & Prais* (1956), *Simon & Bonini* (1958), and *Lydall* (1959).

By way of summarizing these works as well as those appearing later on, one might draw a distinction between the analytical models and simulation models. This is also the dichotomy used in the organization of the present work: Part 2 deals with analytical models, and Part 3 with simulation models.

Another possible distinction is that between models where: (1) the outcomes are jumps between classes; and (2) those where the outcomes are percentage changes in present sizes. These two groups of models are generally referred to as *discrete* and *continuous* models, respectively.[3] This distinction will be made in Parts 2 and 3.

Since most of the earlier models are analytical, Part 2 deals mainly with earlier models, whereas two new models are presented in Part 3. Discussions of the models as well as the results of their application are included in both parts.

The following chapter discusses some concepts related to the research: firm, size, and concentration.

2 Some Basic Concepts

Studies like the present one require the definition of several concepts. Here, the concepts of firm, size, and concentration are of particular relevance. None of them is univocal and many interpretations can be found. However, in empirical applications the choice among alternative concepts depends very much on the data available. This means that quite often a second-best or even third-best solution has to be chosen. For this reason the discussion in this chapter is related to available data sources.

The following section discusses the choice of data sources for the present study. Then the concepts firm, size, and concentration are treated in the next three sections. These sections include general discussions as well as more specific comments on the concepts in relation to the samples of the study.

Data Sources

The fact that size distributions are skewed makes it quite natural to devote the study, at least partially, to the right-hand portion of some distributions. This means that we would be interested in studying the largest firms within some geographical areas. Such data were also easily obtained, as the magazines *Fortune* and *Ekonomen* have published annual lists of the largest firms for certain areas since the mid-1950s.[1]

Fortune has included the 500 largest corporations in the United States and the 200 largest corporations in countries outside the United States (USA*), while *Ekonomen* has listed firms in Scandinavia (SCAN) with a turnover exceeding SCr 100 million (approximately US $20 million). In addition to these three areas (USA, USA*, and SCAN), the largest firms in Europe (EUR) and Sweden (SWED) are studied. These two samples were derived from the lists for USA* and SCAN, respectively. Thus a total of five samples of large firms are studied. The investigation covers the period 1956–65, the period for which data were available for all areas at the time of research.

Some of the models to be discussed deal with more complete size distributions and not just the right-hand portion. Consequently, other data than the material described above were required. A number of relative advantages favored the selection of Sweden as an object of study. While many possible sources of Swedish data were found, only one of them covers entire industry groups and facilitated sampling. This was the register of establishments at the National

Central Bureau of Statistics in Sweden. Among the industries in this register, one expansive and one contractive industry were selected. The basis for this choice was table 2-1, which shows the indices of production for those industries which expanded most and least after World War II.

First, the two extremes in table 2-1, the vehicle and the shoe and leather industries, were chosen. Then the final choice was made with respect to the subgroups in these two industries that represented extremes as to change in production. They were: (1) automobile manufacturing industry—most expanding, see *Sveriges industri,* 1967, p. 166 (CAR); and (2) Shoe industry—least expanding, see *Sveriges industri,* 1967, p. 239 (SHOE).

As for the choice of time period, reorganization in Europe and the Korean War seem to have produced extreme economic conditions up until 1952 (see, e.g., *Lindbeck,* 1968, p. 65 ff.). Consequently, 1952 was chosen as the first year for the study. Since data after 1966 were not available in the data source chosen at the time of the study, the complete period examined was 1952-66.

Finally, it was also considered appropriate to investigate whether similar results might be found in capitalist and noncapitalist countries. Thus samples from some socialist countries were sought. Availability of data was a severe limitation in this context, and consequently, only aggregated data could be used. Moreover, the observations are limited to one year: 1967. For this year data could be acquired from Czechoslovakia, Poland, Rumania, and Yugoslavia.

The data on the largest firms are used in chapters 2, 4, 5, and 8, whereas the data on Swedish establishment distributions are considered in chapters 2, 6, and 9. The socialist data, finally, are used in chapter 6.

Firm[2]

The meaning of the concept "firm" depends to a large extent on the level of description chosen.[3] Moving upwards from a low level of description, we can discuss the firm in terms of the following concepts:

1. Plant
2. Establishment
3. Enterprise
4. Concern (parent company with subsidiaries)
5. Trust and cartel

Movement from the lowest level of description ("plant") in this hierarchy implies decreasing emphasis on technical factors, while progressively more importance is attached to organizational factors and economic power.

The unit of description must be chosen with reference to the purpose of the study in question. There are several problems inherent in making this choice. First of all, there are difficulties in discriminating between the different levels of description. In some cases, units on two different levels may even coincide.

Table 2-1
Indices of Production for Some Industries in 1965 (1946 = 100)[a]

Industry	Index
Vehicle	505
Chemical	414
Iron and metallurgic	396
Clothing	162
Food	152
Shoe and leather	100

Source: *Sveriges industri* (1967, p. 75, table 3).
[a]Other measures of expansion can, of course, be used; but the general pattern is very similar for different measures.

Therefore, we should expect to find doubtful cases irrespective of the level of description chosen. Furthermore, the respective concepts are not clearly defined. "Enterprise," for instance, includes stock corporation, limited partnership, general partnership, and sole proprietorship. Thus, although the concept "enterprise" in most cases refers to a stock corporation, we cannot always be sure of its meaning.

The difficulties in defining the firm concept are even more pronounced as we compare units of different economic systems. Thus Richman regards the concept of industrial enterprise in the Soviet Union as "some hybrid of an American corporation and an American factory" (*Richman*, 1965, p. 53).

The above arguments demonstrate that the firm concept can include many different types of economic units. In this study, we will use three concepts in various contexts: concern (parent company with subsidiaries), enterprise, and establishment. We will use the first concept for the sample of large firms (USA-SWED), "enterprise" for the socialist samples, and "establishment" for the Swedish samples (CAR and SHOE). This choice is a direct result of the data available.

Although it might seem preferable to use only one firm concept throughout the study, this is not necessary as long as we do not try to compare populations containing different economic units.[4]

Size

Discussion

Size usually implies a description in relation to some accepted scale. With respect to firms this description is usually given in relevant economic terms. In this context we can distinguish between (1) measures of input, (2) measures of output, and (3) measures considering both these dimensions.

Examples of input measures are employees, assets, and costs, whereas sales, value-added, and production can be mentioned with respect to output measures. The most common examples of the third group are profit and market value.

With respect to the choice between measures of input and output Bain gives some guidance in a discussion where he concludes:

> In the analysis of firms we are fortunately able in general to base our measures of concentration on output or sales data. In the analysis of plants we are unfortunately forced to accept number of workers employed as the only available common denominator measure of plant sizes and plant concentration." (*Bain*, 1966, p. 9 ff.)

This argument is based on the opinion that the amount of output and sales may be regarded as more direct results of market forces, implying that changes in output occur first, followed by changes in input, i.e., measures of input can be expected to lag in relation to those of output.

Turning to the measures in the third group, market value is a frequently used measure. Market value expresses the stock market's expectations of the firm, and it is defined as the number of shares multiplied by the stock market price. A common topic in the accounting literature (see, e.g., *Robicheck & Myers*, 1965), it is often considered to be a measure of the firm's real value. Among earlier studies of size distributions, *Hart & Prais* (1956) used market value in a study of quoted companies in the United Kingdom. A critical drawback from this measure is that entries and exits of firms are regulated by the Board of the Stock Exchange. This circumstance probably influences the characteristics of the entrants, since firms applying for quotation are carefully investigated with respect to their prospects. Consequently, firms with poor prospects are discouraged from applying or are rejected. This should be kept in mind when examining the properties of quoted companies. If we study the growth and profitability of such companies, as did *Singh & Whittington* (1968), an expected result is that small firms grow more rapidly and are more profitable than larger firms, owing to the reasons mentioned above.

Many of the measures mentioned here are expressed in monetary value. This means that most measures suffer from the drawback that they are influenced by changes in monetary value. Different methods can be used to avoid this problem. The simplest one is presented by *Hart & Prais* (1956, p. 163), who assume that price changes will influence the size of all firms to the same extent. However, this assumption is questionable when studying populations of firms in several different industries or countries. An alternative method is to correct for inflation by means of some index, if an adequate deflator can be found.[5] Another alternative is a method used by Adelman, who required that "the class limit of a given size range represent the same percentage of the industry's total assets in . . . two time periods" (*Adelman*, 1958, p. 898).

Another problem which arises when costs and revenues have to be distributed over time is the valuation of such measures as sales, costs, assets, and profit. The comparison of different firms for different years also presents particular difficulty, and the problem is frequently discussed in economic textbooks (see, e.g., *Davidson et al.,* 1964). Part-time employees can also complicate the valuation of the number of employees.

Most of the measures mentioned here can be expressed in absolute as well as relative terms. In the latter case, size is measured in relation to the total amount of size units in the population in question. The market share of sales is the most common example of this manner of measuring.

Our Measures of Size

The collection of data often implies that a distribution will be truncated. This seems especially to be the case with respect to size distributions of firms, particularly at the lower end, since it is never fully clear when the one-man operation turns into a firm. As a consequence, most data-collecting agencies start their recording at some arbitrarily chosen level. The National Central Bureau of Statistics in Sweden, for example, starts at the level of five employees, others start at four employees, and so forth.

The practice of truncating distributions makes it wise to use the same measure of size as the one forming the basis for the data collection. The fact that a firm, for instance, is in the lowest size-class for one measure does not guarantee that it will be there for another.

The above argument does not mean, of course, that the distribution will be perfect with respect to the measure used in the collection of data, but only that the use of this measure will yield the most accurate distribution for the particular sample. Thus, although this measure gives the best result, improvements may be possible. This will be particularly true in the very lowest classes, where statistics are often highly unreliable, since very few firms entering the lowest class voluntarily report their entry. For the entrepreneur to be included in statistics usually entails a great deal of work, filling in forms, etc., and the motivation to make a new firm known in the census is probably very low.[6] A method of avoiding the statistical problems mentioned is to truncate the distribution at an appropriate size—a solution chosen in the present study.

All the major samples provide information on more than one measure. In accordance with the arguments above, however, the measures used in collecting the data are also used in the analysis. These measures are *sales* with respect to the largest firms, and *number of employees* for the Swedish and socialist samples. Thus we have been able to use a measure of output for the largest firms, but, unfortunately, not for the other samples.

However, if different measures of size should yield similar results with

Table 2-2
Rank Correlation between Different Measures of Size

Measures	Sample	CAR		SHOE	
		1952	1966	1952	1966
Input measures					
Employees—workers		.979	.993	.980	.979
—salaried employees		.965	.785	.852	.856
—costs		.881	.913	.903	.776
Output measures					
Value-added—sales value		.976	.979	.950	.933
Input vs. output measures					
Employees—sales value		.924	.951	.944	.862
—value-added		.932	.961	.922	.860
Costs—sales value		.979	.979	.972	.953
—value-added		.925	.924	.861	.796

respect to the latter distributions, we might not need to worry too much about the available measure of size. For this reason, the relations between the different measures on the Swedish establishment distributions have been investigated. This has been done by means of rank correlation, since ordinary correlation co-efficients do not apply to skewed distributions such as those in question.[7]

Spearman's coefficient of rank correlation has been computed for both the Swedish establishment distributions (CAR as well as SHOE). For these computations the following measures of size were available:

1. *Input measures*
    a. Workers (wage earners)
    b. Salaried employees
    c. Employees (workers and salaried employees)
    d. Costs
2. *Output measures*
    a. Value-added
    b. Sales value

The coefficients have been computed for the first and the last years included in the samples (1952 and 1966). The results obtained are shown in table 2-2. The values in the table show that the different measures of size are highly cor-related. Most of the values are above 0.9, although there also are a few values just below 0.8. As for the correlation figure we are most interested in—that between employees and sales—the results seem to indicate that the use of number of employees instead of sales will not influence the results to any great extent. We should note, however, that the coefficient for SHOE in 1966 is lower than for the other three samples. This is in accordance with the above discussion of

Table 2–3
Some Basic Facts on the Populations of Large Firms

Sample	USA	USA*	EUR	SCAN	SWED
Number of firms	500	200	144	167	112
Size of smallest firm in US $100 million	1.10	2.28	2.30	.20	.20
Size ratio of the largest firm to the smallest	188.48	31.50	31.22	36.86	36.86

input versus output measures, i.e., that declines in input measures are the results of earlier declines in sales. The values obtained should therefore be interpreted as the result of lags in adaptation to a changing commodity market in the period after World War II.

Our results are consistent with those obtained by others. *Rosenbluth* (1955, p. 92) found a rank correlation of 0.958 when he related output and employment concentration. Similar results are reported by *Bates* (1965, p. 133 ff.).

As far as the truncation of the samples is concerned, the distributions of the largest firms are truncated in chapter 4. Prior to truncation the samples had the characteristics shown in table 2–3.

After truncation the five samples had the characteristics shown in table 2–4.

Table 2–4
Some Basic Facts on the Truncated Distributions

Sample	Year	USA	USA*	EUR	SCAN	SWED
Number of firms:	1956	96	25	25	52	42
	1965	169	90	70	121	83
Size of smallest firm in US $100 million	1956	3.74	4.51	4.41	.32	.33
	1965	3.98	5.01	5.01	.40	.40
Size ratio of the largest firm to the smallest	1956	28.87	14.41	14.74	9.39	8.94
	1965	52.10	14.33	14.33	18.43	18.43

Source: The information on the Scandinavian and Swedish firms was taken mainly from *Ekonomen*. However, some information was added with respect to changes over time. Additional information was sought when obvious errors were found in the lists in *Ekonomen*. It was gathered from other sources (e.g., *Svenska Aktiebolag*, 1902–   ) and via personal contacts with the companies.

With respect to the establishment distributions, the lower size limit was set at 20 employees because of the expense involved in the acquisition of these data. Some facts on the samples obtained in 1952 and 1966 are shown in table 2–5.

Table 2-5
Some Basic Facts on the Samples CAR and SHOE

Sample	CAR		SHOE	
Year	1952	1966	1952	1966
Number of firms	37	80	131	91
Average size (number of employees)	277.1	373.9	71.3	68.0

Note: Some very slight differences exist between the data used in this study and those in official publications. But there is no reason whatsoever to believe that the differences lead to any systematic error in the material used in the study. It might also be added that statistical deficiencies are most likely to occur in the very lowest classes. The lower limit chosen here might have eliminated the most uncertain statistics. The primary reason for the slight differences that do exist appears to be problems related to changes in classification (1946, 1952, 1957, and 1959) and the addition of "hidden" units (1954 and 1957). The sizes of the smallest and largest firms may not be disclosed since the primary material is classified.

As for the size distributions from socialist countries, the data refer to industry as a whole in 1967. Some basic information on these distributions is shown in table 2-6.

Period of Observation and Classification

Two further questions of importance with respect to size, and particularly changes in size, are the period of observation and classification.

As for the period of observation, our samples contain annual information on the size variables. Thus a shorter period cannot be chosen unless we are prepared to undertake expensive data collection. It also seems appropriate to

Table 2-6
Some Basic Facts on the Samples from Socialist Countries

Country	Total No. of Enterprises	Source
Czechoslovakia	713	Statistickà ročencka 1968
Poland	49,464	Polish Embassy in Stockholm
Rumania	1,081	Barba (1969)
Yugoslavia	2,467	Velkovlc (1968)

Note: The table indicates that the lowest size considered is smaller in Poland than in the other three countries. However, this will not influence the conclusions as long as we only consider the shapes of the distributions.

consider one year as the longest satisfactory period of observation since the use of longer periods will restrict the number of observations. A restriction in this number is a particular disadvantage when the observations are used in parameter estimations.

Regarding classification, two questions are relevant: (1) What shall the principle of classification be? and (2) How many classes are to be used? With respect to the first question, there are two principal means of classification, namely, that the class limits form either an arithmetic or a geometric series.

The former is useful when the distribution considered is symmetric, while the latter is preferable when analyzing skewed distributions. Given a fixed number of classes, this latter approach will assign more class limits to the part of the distribution where most of the units are gathered. This is illustrated in figure 2-1.

Since the distributions of firm sizes are highly skewed, the geometrical approach is most often used in classifying such data (see, e.g., *Hart & Prais,* 1956, and *Archer & McGuire,* 1965). The law of proportionate effect is also important in this choice. A geometrically ascending classification implies that a jump between adjacent classes represents the same relative growth throughout the entire size range.

The class ranges are usually determined so that the upper limit of a class is twice as large as its lower limit. This method yields a fairly low number of classes for small populations such as the largest firms in an area. This fact brings us to the second question above, i.e., the number of classes to be used. Quotients

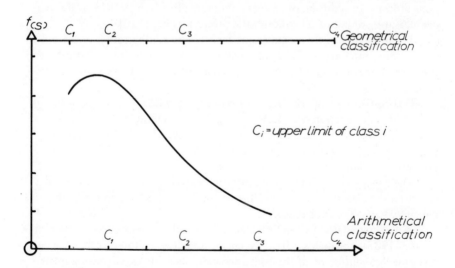

Figure 2-1. Comparison between an Arithmetical and a Geometrical Classification

smaller than two can, of course, be used in the geometrical series in order to obtain more observations. In those cases the following working rules can be used in the choice of quotient (*Wallis & Roberts,* 1956, p. 175):

1.   Between five and fifteen classes is appropriate
2.   The interval should be one-half of the standard deviation of the observations[8]

In our empirical applications we have used the generally applied approach with geometrically rising classes. The quotient used is 2.0 in all chapters except chapter 4, where lower quotients were used in order to attain a larger number of observations.

Concentration

In studying a distribution some summarizing measure is often desirable. The arithmetic mean and variance are frequently used in this context. While these two measures provide useful information about symmetrical distributions, they provide but limited information about skewed distributions. As a consequence, quite a few alternative measures have been suggested in relation to these distributions. Many of them were first suggested in discussions of the distribution of personal income and wealth.[9] Later on, as the issue of industrial concentration grew in importance, these measures were applied to populations of firms. In addition, other measures more directly oriented toward the measuring of industrial concentration were suggested.[10]

The most frequently used measure seems to be concentration ratios, which involve the computation of the joint market share of the *n* largest firms. Several concentration ratios are often derived, such as the ratios for firms having rank 1, 4, 8, and 16.

Difficulties in comparing figures for different industries have given rise to the use of concentration curves. According to this method, a curve of joint market shares is derived as a function of the number of firms. Consequently, CR_n is the height between the curve and the abscissa for *n* firms. Figure 2-2 provides an example of a concentration curve.

The conclusions from the concentration curves are drawn from the order of curves on the diagram. This means that the problem in comparison of industries remains when two curves intersect.

A measure related to concentration curves is the Lorenz curve, presented in *Lorenz* (1905). The major difference between the two methods is that the latter takes the distribution of all firms into account. Such curves will have a shape similar to the one shown in figure 2-3.

From the diagram we can obtain the percentage share of the size units

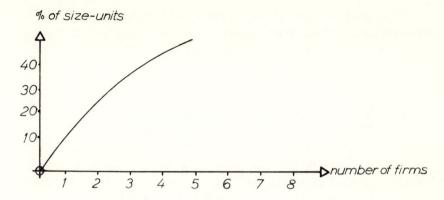

Figure 2-2. Concentration Curve—An Example

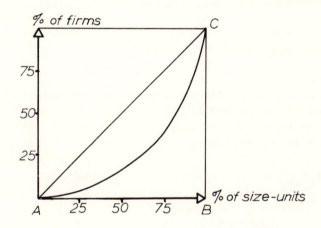

Figure 2-3. Lorenz Curve—An Example

controlled by a certain percentage of firms.[11] Were all firms to have the same size, the Lorenz curve would be equivalent to the diagonal. An increasing degree of concentration thus moves the Lorenz curve from this line. The problem of intersecting curves also exists for this method, but there is a certain advantage in using Lorenz curves since there is a summarizing measure based on the curve. This coefficient, derived by *Gini* (1911), expresses the area between the curve and the diagonal in relation to the area of the whole triangle *ABC*. Thus it will be a value between 0 and 1. A low value of the coefficient indicates a low degree of concentration, and vice versa.[12]

Other measures of concentration based on market shares include Herfindahl's index, Hall & Tideman's index, and entropy. Of these, the first index consists of the sum of the squared market shares (P_i) of all firms and is expressed as follows: [13]

$$HH = \sum_{i=1}^{N} P_i^2 \tag{2.1}$$

The limiting values for this index are also 0 and 1. As for its use, *Adelman* (1969) has criticized the measure for its lack of a test of significance. The same can be said for the index proposed by Hall & Tideman, which is closely related to that of Herfindahl. Hall & Tideman's index (*HT*) can be expressed in the following way:

$$HT = 1/(2 \sum_{i=1}^{N} i P_i - 1) \qquad (0 \leqslant HT \leqslant 1) \tag{2.2}$$

Entropy, finally, was first used in physics and was subsequently employed in information theory (*Shannon & Weaver*, 1949). *Theil* (1967) and *Horowitz & Horowitz* (1968) proposed this measure to express business concentration. It is formulated in the following way:

$$H = - \sum_{i=1}^{N} P_i \log_2 P_i \tag{2.3}$$

The limiting values of the entropy measure are zero and $\log_2 N$. The former value is obtained in a monopoly situation and the latter for a market where all firms have the same market share, i.e., $1/N$. Due to this relation to N, entropy might be less appropriate for use on very small populations.

One means of avoiding the impact of N on the entropy measure is to use a relative measure of entropy proposed by *Horowitz & Horowitz* (1968). This measure is expressed as:

$$HR = H/\log_2 N \tag{2.4}$$

For obvious reasons the limiting values are zero and unity.

In this section we have described a number of methods of summarizing industrial structure. We have argued that the first approach—the use of concentration ratios—involves problems of interpretation since it does not give a summarizing measure for the complete distribution. The other approaches, however, all give such a measure and provide a number of measures to choose from for use in the present study. In order to obtain an impression of the behavior of the measures, we shall apply them to our samples. First we will examine the possibility of discriminating between different areas with respect to concentration, using the samples of large firms in 1965 (USA-SWED). The results obtained are shown in table 2-7. Discrimination between areas is achieved fairly clearly by the Gini coefficient and relative entropy. These measures also yield the same pattern of ranking which, beginning on the left, is 1, 5, 4, 3, 2. The indices of Herfindahl and Hall & Tideman do not provide a clear-cut distinction, since the differences between the values are very small.[14] Similar patterns appear when investigating changes in concentration over time. In this connection we will compare the values of the measures in 1952 and 1966 for the samples CAR and SHOE. The values obtained are shown in table 2-8.

Due to the difficulties in discrimination, it seems appropriate to exclude the Herfindahl and Hall & Tideman indices. Two measures remain. Of these, the Gini coefficient can also be illustrated by the Lorenz curve, which is an advantage. This is especially so since the lognormal distribution lends certain properties

Table 2-7
Values of the Concentration Measures for the Samples of Large Firms

Sample	USA	USA*	EUR	SCAN	SWED
Number of firms	169	90	70	121	83
Gini coefficient	.477	.303	.325	.402	.421
Herfindahl's index	.020	.019	.025	.017	.024
Hall & Tideman's index	.011	.016	.021	.014	.021
Relative entropy	.899	.953	.945	.934	.925

Table 2-8
Values of the Concentration Measures for CAR and SHOE

Population	CAR		SHOE	
Measure	1952	1966	1952	1966
Number of firms	37	80	131	91
Gini coefficient	.709	.801	.438	.391
Herfindahl's index	.120	.125	.017	.020
Hall & Tideman's index	.093	.063	.014	.018
Relative entropy	.719	.636	.924	.939

to these curves. As a consequence, the Gini coefficient and Lorenz curves will be used in the following chapters as alternatives to the measures provided by the stochastic models.

Concluding Remarks

In this chapter we have discussed three important concepts for the study: firm, size, and concentration. With respect to the first concept, we stressed the importance of the purpose of the study and the availability of statistics. The latter circumstance has also been decisive for the firm concept used here. Concepts on three description levels have been used: parent company with subsidiaries, enterprise, and establishment.

Regarding the second concept, it was stressed that the measurement of firm size can be performed in many ways. Our empirical results indicate, however, that different measures yield similar results for a population of firms. The measures used are: sales and number of employees.

In discussing concentration we found measures based on mean and dispersion as well as alternatives to these traditional measures. The alternative measures discussed have employed market shares in one way or another. After having applied some measures to empirical data, the Gini coefficient was chosen as the comparative measure to be used in following chapters.

The purpose of this chapter has been to provide a basis for the analysis of the stochastic models. We will now proceed to these. In so doing we will first discuss analytical models in chapters 3 to 6, followed by simulation models in chapters 7 to 9.

Part 2
Analytical Models

3 Analytical Models in Review

In this second part we will turn to stochastic theories to deal with size distributions of firms. We will here discuss analytical models in general, investigate earlier models and provide a basis for the simulation models in Part 3. In so doing we will use the distinction between models using discrete size and those using continuous size mentioned in chapter 1.

Models Using Discrete Size

In the discrete models changes in size are represented by shifts between size classes. Thus three types of outcomes are possible: a firm can either (1) jump to any of the higher classes, (2) remain in the present class, or (3) drop to any of the lower classes.

In assigning probabilities to these outcomes the discrete models use three assumptions:

1. The probabilities of the different outcomes are constant over time
2. The probability of an outcome is not influenced by earlier outcomes, i.e., the process has no memory
3. All firms are influenced by the same set of probabilities

The first assumption implies that the process does not take into consideration such variations as those caused by the business cycle.[1] The severity of this limitation will depend a great deal on the period to be studied. Thus, if this period is long enough to cover different business conditions, the impact will be less, and vice versa.

The implication of the second assumption is that the probability of a change in size is based only on the present state of the firm, i.e., the earlier pattern of growth is not taken into consideration. This assumption might cause objections since the growth of some firms would appear to be the result of previously high growth rates. However, we should note that some firms fail due to their previous high growth rates, for example, due to financial problems.

The third assumption means that no distinction is made between different types of firms in the industry under study. For example, a firm with highly qualified staff is treated in the same way as one with the less qualified personnel.

Concerning all three assumptions, it should be kept in mind that they simplify the analysis. This circumstance has been discussed by Champernowne, who concludes:

> For gaining insight into the general effect of the probabilities of up-grading and down-grading on the development of a frequency distribution, sweeping simplifying assumptions are nevertheless essential: without them one would never see the wood for the trees. (*Champernowne*, 1969, p. 379)

The probabilities of change in size—the most essential ingredient of the models—can in a discrete model be summarized in a matrix, as is done in figure 3-1.

The figures to the left of this matrix indicate the classes of firm sizes in period t, and the figures above it relate to the classes in period $(t + 1)$. Assume that we want to know the probability of a firm moving from class 2 to class 3 during one period. We then start in row 2 and follow it until we hit column 3, where we find p_{23}. Since all the probabilities in a row are conditioned on the state of the firm, i.e., that the firm is in that particular class, the probabilities of each row add up to 1.0.

The classes 1-5 are easily understood as size classes obtained by some size classification, generally a geometrically rising one. However, the same does not hold for class 0, which is added to the matrix in order to handle entries and exits of firms. Consequently, this class may be considered as a pool

Period $(\underline{t+1})$

	0	1	2	3	4	5	Σ
0	P_{00}	P_{01}	P_{02}	P_{03}	P_{04}	P_{05}	1.00
1	P_{10}	P_{11}	P_{12}	P_{13}	P_{14}	P_{15}	1.00
2	P_{20}	P_{21}	P_{22}	P_{23}	P_{24}	P_{25}	1.00
3	P_{30}	P_{31}	P_{32}	P_{33}	P_{34}	P_{35}	1.00
4	P_{40}	P_{41}	P_{42}	P_{43}	P_{44}	P_{45}	1.00
5	P_{50}	P_{51}	P_{52}	P_{53}	P_{54}	P_{55}	1.00

(Period t)

Figure 3-1. Transition Matrix—An Example

containing potential firms and firms which have failed during earlier periods. Class 0 thus shows the probabilities that a firm from this pool will enter one of the size classes. On the other hand, column 0 gives the probabilities of exit provided that a firm is in a certain class.

The diagonal of the matrix divides probabilities of decline and growth. As a consequence this will be the axis of symmetry when the law of proportionate effect is assumed valid and geometrically rising classes are used. This means, for example, that p_{12}, p_{23}, p_{34}, and p_{45} will have the same size.

A very important effect of the three assumptions mentioned above is that the same matrix can be used throughout the whole process when studying an industry. This in turn means that the changes in size distributions can be analyzed as a finite Markov chain,[2] an approach which Kemeny & Snell describe as: "a stochastic process which moves through a finite number of states, and for which the probability of entering a certain state depends only on the last state occupied." (*Kemeny & Snell,* 1960, p. 207)

As, for example, *Feller* (1960, p. 356) has shown, the final state of a finite Markov chain—the steady state—is unique and independent of the original distribution. In this process the following relation prevails between the distribution in period t and $(t + 1)$:

$$(v_t) [P] = (v_{t+1}), \tag{3.1}$$

where v_t = a vector expressing the distribution in period t

$[P]$ = the transition matrix.

Substituting in (3.1) we obtain:

$$(v_1) [P]^t = (v_{t+1}). \tag{3.2}$$

In steady state the distributions of two consecutive years will be the same. This yields the following relation:

$$u = u [P], \tag{3.3}$$

where u = a vector expressing the distribution in steady state.

Studies using discrete models are *Simon & Bonini* (1958), *Adelman* (1958), *Collins & Preston* (1960–61 and 1961), *Preston & Bell* (1961), *Archer & McGuire* (1965), and *d'Iribarne* (1967). The first of these studies discusses a model assuming the law of proportionate effect. This model is considered in chapter 4. Among studies using discrete models without the law of proportionate effect may be mentioned those by Adelman and Archer & McGuire. Their approach is presented in chapter 5.

Models Using Continuous Size

A model basic to those using continuous size was suggested by *Gibrat* (1931). His model is based on the following two assumptions: (1) that the law of proportionate effect is valid, and (2) that the number of firms is constant. Under these assumptions a lognormal distribution is obtained (see chapter 6), that is, logarithms of size are normally distributed. Such distributions can be characterized by the two parameters, μ and σ, which are the mean and the standard deviation of logarithms.

In addition to lognormality, distributions generated by Gibrat's model will have the following four properties (see, e.g., *Hart*, 1962, p. 30):

1. Average growth is the same in all size classes
2. Variance of growth is the same in all size classes
3. Proportional growth rates are normally distributed
4. Variance of log (size) is continuously increasing

Various studies have investigated these four circumstances. Most of them have focused on the validity of the law of proportionate effect.

In a study of a number of British industries, *Hart* (1962) found no results violating the four properties. Similar findings are also reported in *Archer & McGuire* (1965) and *Hart & Prais* (1956).

As for the second point above, *Morand* (1967) and *Simon* (1964) found no evidence that variance of growth changes with size. The opposite conclusion is reached, however, by *Hymer & Pashigian* (1962 b) and *Singh & Whittington* (1968), who therefore reject the law of proportionate effect. *Samuels* (1965) reached the same conclusion after finding that the growth of very large firms is proportionately larger than that of small firms.

The law of proportionate effect has been discussed not only with respect to internal growth, but also in relation to mergers. Studies of the British brewing industry (*Hart*, 1957) and the largest industrial firms in France (*Morand*, 1967) showed that mergers occurred to some extent in all size classes. A similar finding was reported by *Ijiri & Simon* (1971), whereas *Hart & Prais* (1956) report opposite results.

The fourth consequence is particularly interesting, since it implies an important difference relative to the discrete models discussed in the preceding section. This property implies that the distribution is not in steady state, but that an increasing degree of concentration should be expected.

Kalecki has the following comment on this fourth property:

> Indeed the argument implies that as time goes by, the standard deviation of the logarithm of the variate considered increases continuously. In the

case of many economic phenomena, however, no tendency for such an increase is apparent (for instance in distribution of incomes). (*Kalecki*, 1945, p. 162)

Kalecki's way of attacking the problem is to assume that the standard deviation is constrained by economic forces. In the case of constant concentration this implies that deviation from the mean of logarithms is negatively correlated to size. But Kalecki shows the distribution to be approximately lognormal. He also treats cases where the standard deviation changes and shows that resulting distributions are lognormal under these circumstances, too.

An extension of Kalecki's approach is presented by *Hart & Prais* (1956), who introduce regression toward the mean, a concept used by *Galton* (1892) to explain the stability of men's height. The implication of this regression in Galton's case is that on an average sons will be closer to the mean than their fathers. *Hart & Prais* (1956, p. 172) explain the corresponding movement toward a mean firm size as the result of the efforts of the firms to reach an optimum size. Consequently, the authors argue, firms of greater than optimum size tend to adapt their size downwards, and vice versa.

The models of Kalecki and Hart & Prais both assume the invalidity of the law of proportionate effect. The models described above will be further discussed in chapter 6.

Concluding Remarks

In this chapter we have made a distinction between discrete and continuous models. We might also have discriminated in two other ways: with respect to the use of steady states in the analysis, and with respect to the assumptions as to changes in the number of firms.

However, even if we used either of these latter methods to dichotomize, we would end up with basically the same grouping of models as we have above. This is because the discrete models generally deal with steady states and assume entries, whereas the continuous models do not.

The major difference between the two classes of models with respect to changes in concentration is that the discrete models give information on the expected degree of concentration, whereas the continuous ones only indicate changes.

The following three chapters will discuss the analytical models in more detail, starting with a discrete model that assumes the law of proportionate effect in chapter 4.

4 A Model Using Discrete Size

In our study of analytical models we will start with a discrete model, introduced by *Simon & Bonini* (1958), which assumes the law of proportionate effect to be valid. The model is a continuation of the work presented in *Simon* (1955).

In this chapter the model is first introduced, and some test procedures are examined.[1] The model is then applied to the samples of large firms. Some conclusions are drawn in the final section.

The Model of Simon & Bonini

General Description

The model developed by Simon & Bonini deals with a population of firms arranged in size classes. The following assumptions are made: (1) the law of proportionate effect is valid; and (2) new firms are born at a relatively constant rate in the lowest size class. Given these assumptions, *Simon* (1955) shows that the process moves toward a steady state. The distribution of firms in this state can be approximated by a distribution referred to as the Yule distribution. Simon expresses the frequency function of this distribution as:

$$f(S) = C\,B(S, \rho + 1) \tag{4.1}$$

where

S	is firm size $(S > O)$
ρ	is a parameter
C	is a normalizing constant

$B(S, \rho + 1)$ is the beta function of S and $(\rho + 1)$, which can be written as

$$\int_0^1 \lambda^{S-1} (1 - \lambda)^\rho \, d\lambda = [\Gamma(S)\Gamma(\rho + 1)] / [\Gamma(S + \rho + 1)]$$

The parameter ρ is defined as

31

$$\rho = 1/(1 - \alpha) \tag{4.2}$$

where

α = g/G, taking values between zero and one

g = the net contribution of entering firms, i.e., sizes of entering firms minus sizes of firms that exit

G = the net growth of all firms

For large values of S (4.1) can be approximated by the Pareto distribution having the frequency function: [2]

$$f(S) = M S^{-(\rho + 1)} \tag{4.3}$$

where M is a constant and $S \geqslant 1$.

For practical purposes we need an expression for the proportion of firms above a certain size S. This can be obtained by integrating from S to infinity. The resulting expression is:

$$F(S) = (M/\rho) S^{-\rho} \tag{4.4}$$

which is one minus the distribution function. Using relative sizes, as did Simon & Bonini, it can be shown that $M = \rho$. This means that (4.4) can be rewritten as:

$$F(S) = S^{-\rho} \tag{4.5}$$

The Assumptions

The assumption with respect to the law of proportionate effect is based on studies of cost curves by *Bain* (see, e.g., 1968, p. 175 ff.), who analyzed cost curves of plants and firms and found the average cost curves to be more similar to a horizontal J than to a U, as is generally assumed (see figure 4.1).

As Simon & Bonini point out, this shape of cost curves implies constant returns to scale above a minimum size (S_m). This means that additional input will cause the same proportional addition to output for all sizes above S_m. This question was discussed by *Hymer & Pashigian* (1962 b), who suspect falling average unit costs and, as a result, increasing returns to scale. Their opinion is based on conclusions from one of their earlier studies of industrial structure. According to their findings, no relation exists between the size of a firm and its mean growth rate, but there is a negative correlation between firm size and standard deviation of growth. They argue that these two facts are consistent with constant as well as declining unit costs. Consequently, they judge the last

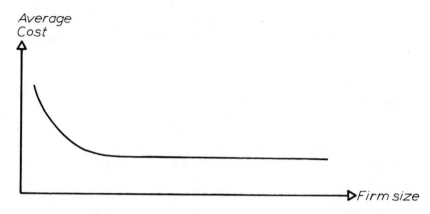

Figure 4-1. Average Cost Curves According to Bain

circumstance to be at least as probable as the first one. *Wedervang* (1965, p. 82 ff.) also doubts this assumption. He expects to find very sharp kinks on the curve at S_m but has been unable to find such kinks. He has instead observed continuous changes, which circumstance he finds more corresponding to a situation involving a slowly falling cost curve than to the type of cost curve assumed by Simon & Bonini.

The second assumption implies that there is a continuous entry of new firms into the lowest class of sizes. This assumption is vital to the shape of the distribution, since it gives rise to the Yule distribution. It implies that entry will vary with the number of existing firms.

The entry assumption has not been discussed to any large extent in earlier studies, but empirical studies indicate that it is quite plausible. Consequently, the model of Hart & Prais presented in chapter 6 of this study has been criticized for *not* assuming entry (see, e.g., *Champernowne*, 1956, p. 182).

In summary, different arguments have been presented concerning the validity of the two assumptions. Consequently, they will be subject to empirical investigation below.

Test Procedures for Simon & Bonini's Model

If a distribution of firm sizes is distributed according to (4.3), the distribution function will appear as a straight line in double logarithmic scale. This method is used as a test procedure by both *Steindl* (1965) and *Wedervang* (1965). In using this method Simon & Bonini state that deviations from linearity should be interpreted as "a reflection of some departure from the law of proportionate effect or from one of the assumptions of the model" (*Simon*

& Bonini, 1958, p. 615). Thus the deviations may be used to locate S_m. Simon
& Bonini warn that the method has not been proven conclusively, however.

  Two other possible interpretations of a systematic deviation from linearity
are:

1. A permament change in the value of α occurs as the number of size units
    increases. This explanation implies that the right-hand side of the straight
    lines is the result of old α-values, whereas the left-hand side expresses the
    more recent values.
2. The model is not valid for the very largest firms, since (4.4) is an approxi-
    mation.

The second interpretation might call for an explanation. (4.4) is obtained by
integrating (4.3) from S to infinity. However, the largest firm does not have an
infinite size. Therefore, if (4.3) is integrated from S to S_{max} (empirically
speaking, the size of the largest firm), (4.6) rather than (4.4) is obtained:

$$F(S) = (M/\rho)S^{-\rho}\left[1 - S/S_{max}{}^{\rho}\right] \qquad (4.6)$$

Where S_{max} is very large relative to S, the latter term will approach zero, but
as S moves toward S_{max}, this will no longer be true. These two interpretations
direct attention to the left-hand side in the event of kinks, while Simon &
Bonini's interpretation, on the other hand, stresses the right-hand side. This
seems to indicate that both sides of a kink ought to be investigated.

  The slope of the line in double logarithmic scale can also be used to obtain
an estimate of ρ. Determination of the parameter ρ makes it possible to perform
a χ^2-test for goodness of fit.[3] Since tests for goodness of fit have their
shortcomings, it seems advisable to apply other tests as complements. One
approach is to look at the parameters. If an empirical size distribution is
Pareto-distributed, α can be derived in several ways. It therefore seems appropriate to make these calculations and to compare the results.

  Let us first look at α in terms of *Simon*'s (1955) analysis of word frequen-
cies in a novel. Here he defines α as the probability that the next word in a text
be a new word. In accordance with his model we make the following assumptions:

1. The size of the smallest firm (S_m) is used as the measure of size, and is
    designated 1 size unit (s.u.)
2. There is a permanent net addition of size units to the distribution.
3. New firms are assumed to have the size 1 s.u.

Letting N denote the number of firms and K the total amount of size units, the
probability that the next added size unit is a new firm can be expressed as:

$$\alpha = dN/dK \tag{4.7}$$

where $0 \leqslant \alpha \leqslant 1$

The above condition implies that the number of firms and the amount of size units must vary in the same direction. Expression (4.7) means that α can be obtained from the slope of a straight line describing the number of firms as a function of all size units. We may, however, suspect that (4.7) will to some extent underestimate α, as entering firms are assumed only to have the size 1 s.u.

A second way of computing α is from transition probabilities. Since these can be estimated, the expected increments from old and new firms can be derived. This in turn makes it possible to obtain an estimate of α, which can be expressed as:

$$\alpha = E(g)/[E(g) + E(G - g)], \tag{4.8}$$

where $E(g)$ = the expected value of g.

Estimates of $E(G - g)$ and $E(g)$ can be derived from the transition matrix, where the class ranges are geometrically ascending. The average size in class i—except in the zeroth class—can then be expressed as: [4]

$$\overline{S}_i = 0.5 \, (k^i + k^{i-1}) = 0.5 \, (k + 1) \, k^{i-1} \tag{4.9}$$

where
$\overline{S}_i$ = the average size in class i

k = the ratio of the upper to the lower class limits in all classes.

A jump from class i to class j will then yield the following change in size units:

$$\Delta = 0.5 \, [(k + 1) \, k^{j-1} - (k + 1) \, k^{i-1}] = 0.5 \, (k + 1) \, (k^{j-1} - k^{i-1}). \tag{4.10}$$

The expected change in size of a firm in class i will then be:

$$E(G)_i = 0.5 \, (k + 1) \sum_{j=1}^{nc} p_{ij} \, (k^{j-1} - k^{i-1}), \tag{4.11}$$

where nc = the number of classes.

The expected total change in size of all firms in class i is obtained by multiplying the expected change in size of one firm by the number of firms in class i (n_i). If all these expected changes in size are added, we obtain:

$$E(G - g) = 0.5 (k + 1) \sum_{j=1}^{nc} \sum_{i=1}^{nc} n_i \, p_{ij} \, (k^{j-1} - k^{i-1}).$$

(4.12)

$E(g)$ is divided into two components, i.e., E(entries) and E(exits). Using the same method as above, we obtain:

$$E(\text{exits}) = 1/2(k + 1) \sum_{i=1}^{nc} n_i \, p_{i0} \, k^{i-1}.$$

(4.13)

E(entries) is estimated as the average annual contribution from new firms. This is due to the fact that the probabilities in the zeroth class cannot be estimated. E(entries) will thus be:

$$E(\text{entries}) = (1/2T) (k + 1) \sum_{j=1}^{nc} n_j \, k^{j-1},$$

(4.14)

where

T = the length of the period observed

n_j = the entries into class j during period t.

Thus we can write

$$E(g) = 1/T \sum_{j=1}^{nc} n_j k^{j-1} - \sum_{i=1}^{nc} n_i p_{i0} \, k^{i-1}$$

(4.15)

$$E(G - g) = \sum_{i=1}^{nc} \sum_{j=1}^{nc} n_i \, p_{ij} \, (k^{j-1} - k^{i-1})$$

(4.16)

Insertion of (4.15) and (4.16) in (4.8) immediately yields an estimate of α.

A third method of estimating α is based on the expression for average size in a Pareto distribution (see *Allen*, 1962, p. 408):

$$\bar{S}_n = S_n \rho/(\rho - 1),$$

(4.17)

where $\bar{S}_n$ = the average size of the n largest firms.

Inserting (4.2) in (4.17) we obtain:

$$\bar{S}_n = S_n/\alpha. \tag{4.18}$$

If we work with relative sizes and assign one size unit to the smallest firm, the average size of all N firms will be:

$$\bar{S}_N = 1/\alpha \ . \tag{4.19}$$

This means that we can write:

$$\bar{S}_N = K/N = 1/\alpha, \tag{4.20}$$

where K = the total amount of size units.

Solving for α we obtain:

$$\alpha = N/K. \tag{4.21}$$

Summarizing, we now have four alternative ways of estimating α. The value of this parameter can be obtained from:

1. dN/dK
2. the transition matrix
3. N/K
4. ρ.

The first two methods are related to Simon & Bonini's interpretation of α, while the latter two are linked to the use of the Pareto distribution. Simon & Bonini use g/G for a single year as the expression for α. But this does not seem to be a good estimate compared to the α obtained from the Pareto distribution, since g/G can be expected to vary a great deal between different years.

A further check on the appropriateness of the model is to look at the value of the parameter ρ. Since α takes values between zero and unity, ρ must be equal to or larger than unity.

An additional check of the model is based on the relation between α and the Gini coefficient as derived by *Quandt* (1966 b). He shows that for a Pareto distribution the area to the left of the Lorenz curve can be written as (*ibid.*, p. 62):

$$D = \rho/(2\rho - 1) = 1/(1 + \alpha). \tag{4.22}$$

This means that the Gini coefficient can be written as:

$$R = (1 - \alpha)/(1 + \alpha). \tag{4.23}$$

The test of the model should also include tests of the validity of the assumptions. *Friedman* (1953) has argued that a model should not be valued on its assumptions but on its predictive value. In many cases, however, it is difficult to judge how "good" a prediction really is. This is particularly true for the kind of problems we are dealing with here since the techniques for discriminating between distributions are often inadequate. Consequently, as many tests as possible should be applied to our models. Moreover, doubts as to the validity of the assumptions have been expressed, as mentioned earlier.

With respect to the law of proportionate effect, *Simon & Bonini* (1958, p. 612) propose two testing methods:

1. Plotting the firm sizes at the beginning and the end of a period on a double-logarithmic scale. The law can be accepted if the slope of the regression line is 45°, and the plots homeoscedastic.
2. By estimating probabilities and testing whether the probabilities of relative growth rates are equal.

Simon & Bonini used these two methods and report results supporting the validity of the law when applied to the largest firms in the United States, 1955-56. A *t*-test may be used with reference to the first of their methods. In this test the coefficient of regression is tested for deviation from unity.

As for the second assumption, no test methods have been proposed in the literature. This is probably because the number of observations is usually limited. The only possible check would seem to be direct observations as to whether or not entries occur.

Before applying the model to empirical data, we will also discuss its applicability to aggregated data. The question has been raised by *Quandt* (1966 a), who strongly opposes the use of aggregated data because of the great inequalities between different industries. *Simon & Bonini* (1958, p. 612), on the other hand, give the following arguments for using aggregated data: (1) the model in no way refers to cost curves of sample firms;[5] and (2) if firms within different industries are distributed according to the Pareto curve, the composite curve will also have this appearance.

Quandt claims that both these arguments constitute a very weak defense of aggregate analysis. His argument is interesting in view of the results of tests he performed, however. Quandt used data on the 500 largest firms in the USA as listed in *Fortune* (1955 and 1960) as a composite sample and made samples for thirty industries according to SIC.[6] In his study six distributions are presented, among them three distributions advanced by Pareto.[7] Quandt tested the goodness of fit with four kinds of tests and obtained results which can be summarized as follows:

1. The Pareto distributions were rejected in 6 cases out of 24 when applied to

Fortune data. The corresponding figures for the other three distributions
were 22 out of 24.[8]
2.   The Pareto distributions provided a very poor explanation of the size
     distributions within industries. These distributions were rejected in 130
     cases out of 360, compared to 47 cases out of 360 for the other three.

A probable interpretation of these results is that the number of firms above S_m
within industries is too low for the Pareto distribution to be an acceptable
description. Steindl has argued for this interpretation in the following way:

> If the mass of firms is divided according to individual lines of manufacturing
> or of trade, the distributions often become irregular. A neat division of
> firms, if it goes beyond the broad divisions of manufacturing, trade, etc., is
> artificial, because of the arbitrary allocation of many firms, and because
> firms, in growing, spread from one line of business to the other; the
> stochastic process which accounts for the regularity is more applicable to
> the broad field of all firms than to narrow industrial divisions. (*Steindl*,
> 1965, p. 194)

In accordance with this interpretation we will in the following empirical section
investigate only the samples of large firms (USA–SWED). They will be examined
by means of the methods described in this section.

Empirical Section

Test of the Model

   In investigating Simon & Bonini's model in relation to the five samples of
large firms, we have chosen the limits of size classes in order to compose a
geometric series, as was discussed in chapter 2. The ratio (k) between the upper
and lower class limits is determined so as to establish ten classes.[9]
   As a first test the observations are plotted in double logarithmic scale. The
resulting diagrams all resemble the one shown in figure 4–2, which refers to
SWED. All indicate kinks occurring on the left-hand side of the curve.
   The least-square estimates of ρ and $A = \log(M/\rho)$ are computed, as shown
in table 4–1, and are used when testing for goodness of fit by a χ^2-test. Ten
classes were used in this test, which gives a critical value of 18.48 at the 1
percent level of risk (7 d.f.).[10]
   Since all χ^2-values are larger than the critical value, none of the fits is
acceptable on the chosen level of risk. This indicates some deviations from
linearity. Another hint is that the values of A all exceed zero. It is difficult to
judge where the deviations start. But according to the method used by Simon

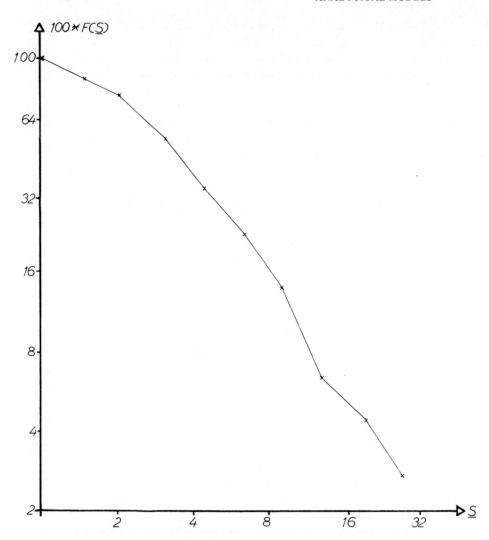

Figure 4-2. Proportion of Firms above a Certain Size among the 112 Largest Swedish Industrial Corporations (Double Logarithmic Scale to the Base of 10)

& Bonini, the approximate locations can be estimated, with results as shown in table 4-2.

In compliance with the previous discussion of deviations from linearity, the regression coefficients on both sides of the kink are computed. The upper (left) part of the lines yields the results shown in table 4-3.

Table 4-1
Results of the Estimations and Tests

Sample	USA	USA*	EUR	SCAN	SWED
ρ	1.31	1.84	1.71	1.32	1.18
A	.20	.49	.44	.18	.17
χ^2-value	84.30	46.30	37.24	31.57	21.82

Note: $A = {}^{10}\log(M/\rho)$.

Table 4-2
Facts about the Kinks in the Double Logarithmic Diagrams

Sample	USA	USA*	EUR	SCAN	SWED
Location of kink in $100 million	3.98	5.01	5.01	.40	.40
Number of firms above the kink	169	90	70	121	83

Table 4-3
Results of the Analysis of the Left Side of the Five Distributions

Sample	USA	USA*	EUR	SCAN	SWED
ρ	.83	.98	.92	.52	.48
A	.01	.00	.00	.00	.00

None of the values of ρ meets the condition of being larger than unity, which means that the model is unacceptable for the left-hand side of the distribution.

Turning to the right-hand side of the distribution, we obtain the results shown in table 4-4. The test for goodness of fit is again performed on the 1 percent level of risk.[11]

None of the differences exceeds the critical value on a 1 percent level of risk. Thus the Pareto distribution is not rejected as a description of the size distribution of the largest firms in the five areas. As a further check of the distribution, the parameter α is computed using the five methods described in the earlier section, "Test Procedures for Simon & Bonini's Model." These calculations yield the results shown in table 4-5.

The following conclusions may be drawn from table 4-5:

Table 4–4
Results of the Analysis of the Right Side of the Five Distributions

Sample	USA	USA*	EUR	SCAN	SWED
ρ	1.44	1.97	1.85	1.62	1.46
A	.10	.01	.04	.10	.12
χ^2-value	15.82	11.60	4.59	18.16	6.33
Critical value	18.48	15.09	11.34	18.48	15.09
Degrees of freedom	7	5	3	7	5

Table 4–5
Values of α Obtained when Applying Different Methods of Estimation

Sample Estimate from	USA	USA*	EUR	SCAN	SWED
a. Matrix	.318	.592	.595	.334	.260
b. dN/dK	.248	.516	.450	.315	.243
Average (a. and b.)	.283	.554	.523	.325	.252
c. g/G	.247	.223	.276	.337	.217
d. ρ	.306	.492	.459	.383	.315
e. N/K	.293	.482	.444	.368	.314
Average (d. and e.)	.300	.487	.452	.376	.315

1. Values obtained using dN/dK are all lower than those derived from the matrix. This is probably because the dN/dK method assumes the size of new firms to be 1 s.u. (see earlier section, "Test Procedures for Simon & Bonini's Model").
2. g/G appears to be an acceptable estimate of α. Note, however, the low figures for USA* and EUR, which indicate that this estimate has short-comings.
3. The values derived from ρ and N/K, both of which are based on the assumption of a Pareto distribution, show a close similarity. This can be interpreted as further support for the model, since the distributions investigated satisfy the expression for average size in a Pareto distribution fairly well. All values from ρ exceed the values from N/K.
4. No real pattern can be observed with respect to the two averages. The first average gives the largest values in some cases, the second in others.

Table 4-6
Values of the Gini Coefficient

Sample	USA	USA*	EUR	SCAN	SWED
Value of ρ	1.44	1.97	1.85	1.62	1.46
Gini coefficient from Yntema's formula (R_Y)	.477	.303	.325	.402	.421
Gini coefficient from ρ and actual value of $M(R_{\rho 1})$	.722	.354	.439	.623	.736
Difference from R_Y	.245	.051	.114	.221	.315
Gini coefficient for $M = \rho(R_{\rho 2})$	.532	.340	.370	.446	.521
Difference from R_Y	.055	.037	.045	.044	.100

Turning, then, to measures of concentration, values of the Gini coefficients are computed. This is accomplished by a formula suggested by *Yntema* (1933, p. 428) as well as from ρ. In the latter case, both actual values of M and $M = \rho$ are used. The Gini coefficients are shown in table 4-6, and the Lorenz curves in figure 4-3.

A comparison of the different R-values shows all R_ρ-values to exceed R_Y-values.[12] The differences are smaller, however, when $R_{\rho 2}$-values are considered. The results imply that concentration is overestimated when ρ is used in estimates of the Gini coefficient.

With respect to ranking, the values of the Gini coefficient exhibit the same pattern as those for ρ, with one exception: with regard to $R_{\rho 1}$, SWED and USA change places, probably due to the large A-values for SWED (see table 4-4).

Unfortunately, we have no test statistics for the values of the Gini coefficient, and we can only note that the values obtained seem reasonable. Thus we still have no reason to reject the model for the truncated samples. A further test remains, however: a test of the assumptions.

Testing the Assumptions

In testing the law of proportionate effect, data from 1964 and 1965 are plotted in double logarithmic scale (see earlier section, "Test Procedures for

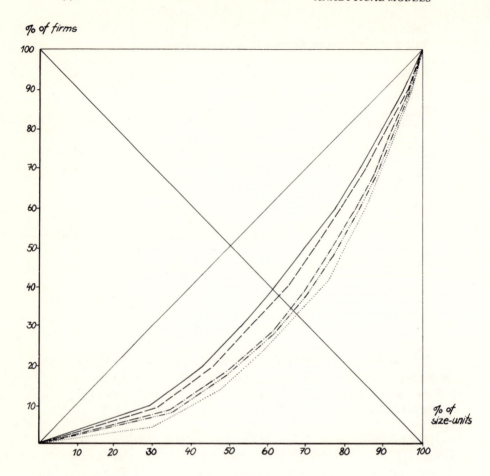

Figure 4-3. Lorenz Curves for the Five Samples of Large Firms (USA–SWED)

The curves represent populations investigated in
USA* ————
EUR ————
SCAN —·—·—
SWED —···—
USA ·········

Simon & Bonini's Model"). This has been done for all five samples, as in the diagram for SWED shown in figure 4-4.

The conditions of homeoscedasticy and slope appear to have been met in all the diagrams. Regression coefficients were computed for the five samples with reference to the slope. They have been tested for deviations from $\beta = 1.0$ with a

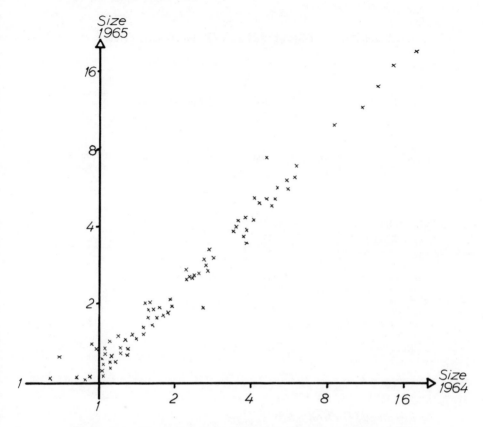

Figure 4-4. Sizes of Firms Investigated in Sweden: 1964 and 1965 (Double Logarithmic Scale to the Base of 10)

t-test on the 1 percent level of risk. The results of the computation and the test are shown in table 4-7.

The values obtained from our tests using the first method show that the law of proportionate effect may not be rejected for the five samples. The second method of testing the law will be considered in the following chapter, where transition probabilities are estimated.

A study of the distributions during the period 1956–65 indicates that entries of new firms take place. Some data on the net entry into the five samples is shown in table 4-8. The relatively few observations do not allow us to establish any trends with respect to entry. Thus the assumed relation to the existing firms has not been verified.

In conclusion, we can note that none of our tests has indicated that the model of Simon & Bonini should be rejected for the truncated samples. These results will now be discussed in relation to those obtained in earlier studies.

Table 4–7
Values Obtained in the Tests of the Law of Proportionate Effect

Sample	USA	USA*	EUR	SCAN	SWED
Regression coefficient (b)	1.005	.972	.966	1.016	1.029
Standard deviation (s_b)	.014	.015	.015	.028	.035
Degrees of freedom	146	81	62	84	76
t-value	.346	−1.860	−2.155	.580	.812
Critical value	2.61	2.65	2.66	2.65	2.65

Table 4–8
Information on Entry

Sample	USA	USA*	EUR	SCAN	SWED
Net increase in number	73	65	45	52	42
Mean entry/year	8.1	7.2	5.0	7.0	3.9
Largest value	16	12	10	12	6
Smallest value	−1	−5	−3	0	0

Our Results and Earlier Studies

The model has previously been applied to data by *Simon & Bonini* (1958), *Steindl* (1965), *Wedervang* (1965) and *Quandt* (1966 a).

Simon & Bonini applied the method described to data on the 500 largest industrial firms in the United States in 1955, and to data on British firms collected by Hart & Prais. They report good fits and values of α similar to g/G in both cases, although they note in a footnote: "In the absence of better developed theories about goodness of fit of these skew distributions than we now have, we prefer not to make definite statements about 'how good' the fits are" (*Simon & Bonini*, 1958, p. 612 n.). If we apply the same method as the one used above to the population of American firms in 1955, deviations from linearity can be found. A χ^2-test also yields a significant result.[13] If S_m for 1965 is deflated and used as S_m for 1955, the test does not reject the fit to the Pareto distribution for firms above this size.[14] The analysis is summarized in table 4–9.

The table indicates some doubt as to whether the nontruncated distribution used by *Simon & Bonini* (1958) follows the Pareto distribution.

Wedervang has applied the method to Norwegian firms in 1930 and 1948. He is not satisfied with the fit and writes,

Table 4–9

Results of the Analysis of the Largest American Firms in 1955

Number of firms	S_m[a]	ρ	A	χ^2	g/G	$\alpha_{N/K}$
500[b]	.55	1.24	.18	82.44	.212	.279
91	3.63	1.42	.06	9.52	n.a.[c]	.314

[a]In US \$100 million.

[b]These figures are derived by me. Simon & Bonini report ρ = 1.23, which corresponds to α = 0.187.

[c]Not available.

> If the distribution follows the Pareto law, the curves should be straight lines. This is evidently not true of any of the curves, except for the higher ranges of added value (kr. 1–20 million) where the curve is approximately straight with a Pareto coefficient of roughly –1.4. (*Wedervang*, 1965, p. 82)

The curves shown in Wedervang's study have the same general appearance as those obtained in the present study. This is probably because the supposed J-shape of the cost curves implies a successive change from falling unit costs to constant unit costs. Wedervang's rough measure of ρ = 1.4 [15] corresponds to a value of α = 0.286. This can be compared with $\alpha_{N/K}$, which is 0.345.

Steindl also reports deviations from linearity and states: "The Pareto law is valid only for a very small percentage of firms, but they are the firms which really account for the bulk of economic activity" (*Steindl*, 1965, p. 187). He considers the most common values of ρ to be in the range from 1.0 to 1.5. His investigations of American and German firms give values of ρ which are 1.1 and 1.3, respectively.

Quandt (1966 a) also tested the fit of the Pareto distribution to *Fortune* data. He does not give any values of ρ, but reports fits which have to be rejected only a few times (see the section, "Test Procedures for Simon and Bonini's Model"). This is especially interesting since Quandt's analysis concerns 500 firms in 1955 and 1960, which means that his sample contains quite a few firms which, according to our discussion above, are far below S_m.

The various techniques for testing distributions are topics of considerable debate. Test results are to some extent linked to the technique applied. Nevertheless, for both samples used by Quandt, we can observe deviations from linearity on the left-hand side similar to those in our own analysis.

Conclusions

The main conclusion drawn in this chapter is that Simon & Bonini's model cannot be rejected for the distributions of very large firms. Despite its limitation with respect to applicability, the model seems to have considerable value, particularly in view of the high ratio of production originating from the largest firms in an area.

We have observed systematic deviations from linearity as size decreases. This holds true for the samples in 1965 as well as for Simon & Bonini's data. Regression lines for sizes below S_m have regression coefficients below unity, an interesting fact which may be interpreted to support the assumption of constant unit cost curves above a certain size S_m. But it is very difficult to estimate this size, and the values above have been taken from diagrams. With respect to these estimates we observe that the values for USA, USA*, and EUR do not differ very much, whereas the values for SWED and SCAN are about one-tenth of these. To some extent the differences might depend on errors in the estimates of S_m. We suspect this mainly because A for USA, SCAN, and SWED exceeds zero. This is a likely explanation for the difference between USA on the one hand and USA* and EUR on the other. It might also provide a partial explanation with reference to S_m for SCAN and SWED, although the whole difference cannot possibly be explained by this factor alone. Other possible explanatory factors include a weaker state of competition and more advanced specialization among firms within SCAN and SWED.

The ratios of concentration, however, show nearly the same levels for SWED and USA, which could be cause to reject at least the first of these inter-pretations. But, since the firms of the aggregates used compete in very different markets, this may be a false conclusion.

The regularity in instances where areas with low ρ-values also have a high degree of concentration is noteworthy with respect to the ratios of concentra-tion. This is consistent with the definitions. A low value for ρ means a slight increment from new firms, which is usual in areas with a high degree of con-centration.

Another interesting fact is the similarity between ρ-values for American firms in 1955 and 1965 when the same S_m-limit is used. The link between ρ and the Gini coefficient leads us to suspect that the degree of concentration above S_m is about the same in 1955 and 1965. This is in accordance with findings by *Adelman* (1964, p. 231).

As for the different methods of deriving α, we may conclude that values from ρ and from the formula for average size in a Pareto distribution are very close. This can be regarded as further support for the model.

We did not treat the transition process in this chapter, but discussed only the resulting steady-state distribution. In order to find out more about the changes in size distributions of firms, we will examine the transition process in chapter 5. The approach there will also make it possible to discard the assump-tion of the validity of the law of proportionate effect.

5 The Transition Process

In this chapter we are going to examine the process behind the steady-state distributions more closely, discussing and analyzing the transition process in terms of a Markov chain.[1]

Methods of estimating transition probabilities are discussed in the next section. Such estimates make it possible to derive steady-state distributions, the topic of the following section. The transition probabilities and the steady-state distribution may be used to compute an index of mobility formulated by *Adelman* (1958). Her measure as well as some other methods are described under the subheading, "Measures of Mobility." In an empirical section the techniques described in earlier sections are applied to the samples of large firms (USA–SWED). Transition probabilities and steady-state distributions are derived. The latter are then compared with actual distributions. Finally, the values from two indices of mobility are compared.

Estimating Transition Probabilities

Transition probabilities may be estimated if we assume that the matrix they constitute is constant over time. Such an assumption may arouse objections due to suspicions that the probabilities differ during different phases of the business cycle. An approach which avoids such criticism is to choose the estimation period so that it covers various business conditions. This may be easier said than done, since it is difficult to state the definite length of the cycle. However, we know from earlier research that the average length of business cycles is 4–5 years in the United States and 7–9 years in the United Kingdom.[2]

Adelman (1958) and *Archer & McGuire* (1965) investigated samples of American firms, using periods of two decades (1929–39 and 1945–56) and fourteen years (1947–61), respectively. Although Archer & McGuire use a longer period than those used by Adelman, they consider a fourteen-year period to be "shorter than ideally desirable" (*Archer & McGuire*, 1965, p. 235). But they do consider the period chosen "long enough to permit the conclusions to have some value since change is investigated over a variety of business conditions" (*Ibid.*, p. 235). The appropriate length of an observation period is obviously also related to the size of the population used, since this influences the number of observations underlying the estimates of the probabilities. In Adelman's case the number of observations was about 2,100.

We will here consider two methods for the estimation of transition probabilities: (1) the maximum-likelihood method, and (2) the least-square method.

The actual transitions provide the bases for the estimates according to the first method. If we start by assuming that the law of proportionate effect is valid, the estimations are carried out as follows.

1. The actual changes in size during periods of time are observed.[3]
2. The probabilities are estimated as:[4]

$$p_{i,i+1}^* = \sum_{i=1}^{nc-1} \sum_{t=1}^{T} a_{i,i+1,t} / \sum_{i=1}^{nc} \sum_{j=0}^{nc} \sum_{t=1}^{T} a_{ijt} \qquad (5.1)$$

$$p_{ii}^* = \sum_{i=1}^{nc} \sum_{t=1}^{T} a_{iit} / \sum_{i=1}^{nc} \sum_{j=0}^{nc} \sum_{t=1}^{T} a_{ijt} \qquad (5.2)$$

$$p_{i,i-1}^* = \sum_{i=1}^{nc} \sum_{t=1}^{T} a_{i,i-1,t} / \sum_{i=1}^{nc} \sum_{j=0}^{nc} \sum_{t=1}^{T} a_{ijt} \qquad (5.3)$$

where

p_{ij} = the probability of moving from class i to class j

a_{ijt} = the number of movements from class i to class j during period t

nc = the number of classes

T = length of the period used for estimation.

In this instance entries cannot be included in the estimates, since the number of potential entrants (box 00 in the matrix) is unknown. This problem can be dealt with in two ways:

1. The probability estimates obtained for nonzero classes are also supposed to be valid for the zeroth class. Thus we assume that the probability of staying $= p_{ii} + p_{i,i-1}$.
2. Entries are assumed to be governed by a special process.

The problem concerning entries does not arise if the law of proportionate effect is dropped. Instead, an arbitrary number can be chosen which expresses the sum of events in the zeroth class. The figure chosen does not influence the solution as long as the matrix is used to derive the steady-state distribution, as was shown by *Adelman* (1958, p. 901, n. 16). She assumed this figure to be 100,000 and motivates her choice as follows:

Our choice of number was guided by the desire to keep the reservoir of incipient enterprises large by comparison with the number of corporations actually within the industry. But this arbitrary selection does not affect the economically relevant portion of our results. (*ibid.* p. 899)

In this case, equations (5.1–5.3) can be replaced by:

$$p_{ij}^* = \sum_{t=1}^{T} a_{ijt} / \sum_{j=0}^{nc} \sum_{t=1}^{T} a_{ijt}. \qquad (5.4)$$

It should be noted that the estimates derived by (5.4) are less safe than estimates from (5.1–5.3) when applied to the same total amount of data. This fact can cause difficulties when the assumption is discarded.

A description of the least-square method is to be found in *Telser* (1963). This approach is particularly useful when the actual transitions are not known. It implies that actual distributions are used in a system of regression equations so as to obtain least-square estimates of the probabilities. These equations have the following basic form:

$$m_{jt} = \sum_{i} m_{i,t-1} p_{ij} + v_{jt}, \qquad (5.5)$$

where

p_{ij} = the probability of transition from state i to state j

v_{jt} = a random variable; $E(v_{jt}) = 0$

m_{jt} = the proportion of units in state j at period t.

Telser points out that this equation

suggests that we may be able to estimate the p_{ij}, given readings on m_{jt} and m_{it-1} ($i = 1, \ldots, r$) even if we do not know the actual number of transitions n_{ij}. We can regard [5.5] as a linear regression of m_{jt} on $m_{i,t-1}$ in which the coefficients of $m_{i,t-1}$, the transition probabilities, are to be estimated by least squares. (*Telser*, 1963, p. 277)

As Telser points out, the least-square estimates will not be minimum-variance linear unbiased estimates since $m_{i,t-1}$ are stochastic variables. Another circumstance of importance in the present context is the difficulty in including the zeroth class. A further drawback is that the least-square technique may yield estimates which are negative or greater than one.

As for this last circumstance, *Telser* (1963, p. 279) suggests that inconsistent estimates be replaced by zero or one. A more accurate method is the application of mathematical programming. Charnes & Cooper express the idea behind this approach to estimation in the following way:

> Sometimes, however, management-planning problems assume forms that make standard statistical techniques inapplicable without further adaptation. This is not merely a matter of the size and complexity of these problems; other features may also demand attention. For instance, some information may be known with perfect certainty in advance, so that it is not desirable to let statistical fluctuations alone decide the qualitative behavior of all parameters. This is especially true of bounds, or limits, which may be stated in the form of inequalities. (*Charnes & Cooper*, 1961, p. 327)

Thus, if we know certain bounds and limits of a parameter, this technique will make it possible to improve the estimates obtained from standard statistical techniques.

Charnes & Cooper suggest that the nonlinear problem is transformed into a linear one. An alternative—quadratic programming—is suggested by *Theil & Rey* (1966).

The least-square method is particularly convenient when actual transitions are not known or when the population under study is of extreme size. Circumstances such as these seldom arise when size distributions in single industries are studied. Consequently, earlier studies have used the maximum-likelihood method (see *Adelman*, 1958, and *Archer & McGuire*, 1965). This is also the method to be applied below.

Steady-State Distributions

In chapter 3 we described the relation between distributions in two consecutive years when a steady state is attained. The relation is (see equation 3.3 above):

$$u = u\,[P] \qquad\qquad\qquad\qquad (5.6)$$

where

u = a vector expressing the steady-state distribution,

$[P]$ = the transition matrix.

(5.6) immediately produces a system of equations with as many equations as unknowns. However, since the equations in the system are linearly dependent one of them has to be excluded. The missing equation can then be obtained from the relation:

$$\sum_{i=1}^{nc} u_i = 1.00, \qquad (5.7)$$

where

u_i = the ith element of the vector u,

nc = the number of size classes.

The system now contains the desired number of independent equations and is soluble. The solution is the steady-state distribution which also includes the zeroth class. But this is only a means of solution. When the zeroth class is eliminated, a distribution comparable to actual ones is obtained.

An alternative method of deriving the steady-state distribution is presented by *Champernowne* (1969, p. 380). His approach is based on the fact that in steady state the inflow and outflow of any given class will be equal. This means that equations of the following form can be derived. [5]

$$\begin{cases} p_{01} u_0 = p_{10} u_1 & (5.8) \\ p_{12} u_1 = p_{21} u_2 \\ \quad . \\ \quad . \\ \quad . \\ p_{n-1,n} u_{n-1} = p_{n,n-1} u_n \end{cases}$$

Rewriting the system of equations, we obtain:

$$\begin{cases} u_1/u_0 = p_{01}/p_{10} & (5.9) \\ u_2/u_1 = p_{12}/p_{21} \\ \quad . \\ \quad . \\ \quad . \\ u_n/u_{n-1} = p_{n-1,n}/p_{n,n-1} \end{cases}$$

This system of equations also yields the steady-state distribution when combined with (5.7).

A stationary as well as a moving equilibrium can be derived by the methods described. These two types of equilibria differ in their definition of the class ranges. In the stationary analysis the ranges are expressed in absolute terms; in the moving analysis they are related to the total amount of size units. *Adelman* (1958) and *Archer & McGuire* (1965) derived and explained stationary equilibria. An example of the second approach is the model described in chapter 4.

One of the most important advantages of the approach involving moving equilibria is that the analysis can be extended to sizes which are much larger than the largest firms of today. This applies to the stationary approach only when the law of proportionate effect is assumed valid. Otherwise, the analysis is limited to existing size classes.

The main disadvantage in using moving equilibria is that they require a great deal of computational work. In addition, it is difficult to interpret the relative sizes. These two disadvantages highly complicate the derivation of steady states for moving equilibria.

Assumption of the law of proportionate effect can also cause problems with respect to steady states. *Champernowne* (1969, p. 390) shows that no steady state may be obtained if the law of proportionate effect is assumed valid when an unlimited number of classes is used.

If all firms start in class zero, the frequency distribution at time t will have

$$\text{mean} = t(p_{i,i+1} - p_{i,i-1}) \text{ and}$$
$$\text{variance} = t[p_{i,i+1} + p_{i,i-1} - (p_{i,i+1} + p_{i,i-1})^2].$$

The parameters of the distribution will undergo a steady increase. A slight modification can transform this process into one involving a steady state. As described by Champernowne: the probability of stepping down from the lowest class is assumed to be zero. As a result, the probability of remaining is assigned the value $(p_{i,i-1} + p_{ii})$ in class 1. But a condition for a steady state is that $p_{i,i-1} > p_{i,i+1}$. The steady-state distribution is thus:

$$\frac{p_{i,i-1}}{p_{i,i-1} - p_{i,i+1}} \quad \frac{p_{i,i+1}}{p_{i,i-1} - p_{i,i+1}}, \ldots, \frac{(p_{i,i+1})^{n-2}}{(p_{i,i-1})^{n-2}(p_{i,i-1} - p_{i,i+1})}. \quad (5.10)$$

This approach could not be used in the applications below since the probabilities of growth were larger than those of decline for all areas except USA*. Thus, we have used the method described in equations (5.6) and (5.7). In dealing with the law of proportionate effect the number of classes was restricted.

Measures of Mobility

The matrix can also be used to estimate the average time spent in the different classes. An expression for the average time spent in a social class was first developed by *Prais* (1955, p. 59). *Adelman* (1958, p. 897) later applied it to industrial firms. Her application leads to the following expression for the average time spent in the ith class of the size distribution of firms:

$$t_i^a = 1/(1 - p_{ii}) \ [6]$$ (5.11)

Similarly, standard deviation for time is expressed as:

$$\sigma_i^a = \sqrt{p_{ii}} \ / (1 - p_{ii}) = t_i^a \sqrt{p_{ii}} \ ,$$ (5.12)

where

   p_{ii} = the probability of a firm remaining in class i in the next period

   t_i^a = average time spent in class i in the actual distribution

   σ_i^a = the standard deviation for time.

The derived average time spent in the ith class can then be compared with the average time spent in classes in an industry with perfect mobility. In this hypothetical industry the probability of a step from class i to class j must be independent of i. This means that the columns of the matrix contain i probabilities of equal size. If we require the actual and the perfectly mobile industry to have the same shape in steady state, then the matrix for the industry with perfect mobility can be derived from the steady-state distribution. This operation yields $p_{ii} = u_i$. As before, the average time spent in class i in the industry with perfect mobility is derived as follows:

$$t_i^m = 1/(1 - u_i).$$ (5.13)

Prais (1955) compared t_i^a and t_i^m in every class. *Adelman* (1958) developed the method further and derived an index for industrial mobility. This index is the ratio of the time spent by all firms in the perfectly mobile industry to the corresponding figures for the actual industry. In terms of our notations, this can be expressed as:

$$M_t = [\sum_{i=1}^{nc} u_i/(1 - u_i)] / [\sum_{i=1}^{nc} v_{it}/(1 - p_{ii})] ,$$ (5.14)

where

   M_t = the index of mobility at time t,

   v_{it} = the ith element of the vector v_t.

The index can be simplified when the law of proportionate effect is valid. In this case p_{ii} is the same for all classes. Since $\sum\limits_{i=1}^{nc} v_{it} = 1.0$ we can write:

$$M_t = (1 - p_{ii}) \sum_{i=1}^{nc} u_i/(1 - u_i). \tag{5.15}$$

The maximum of M_t is 1.0, which is obtained in the case of an industry with a mobility equal to that of the perfectly mobile industry. In Adelman's study the indices reported were about 0.10.

It should be kept in mind, however, that the industry with perfect mobility used in the index is very unrealistic. We should also remember that the index was first derived in order to express social mobility. Transitions between extreme classes in this area can be expected to exist, although the probability of this event is very small. With reference to firms, however, steps over the whole spectrum of classes should be regarded as almost impossible.[7]

Adelman's index is not the only method that can be used to express mobility. Mobility is usually analyzed by observing changes in measures of concentration (see chapter 3). Studies using this approach include those by *Adelman* (1951), *Rosenbluth* (1954), *Einhorn* (1962) and *Carling* (1968).[8]

Special measures of mobility have also been presented. Studies by *Prais* (1958), *Joskow* (1960) and *Hymer & Pashigian* (1962 a) involve three different measures which do not use transition probabilities.

The first two measures have one major drawback. The same number of firms has to be observed in both years. Since changes in this number influence the degree of concentration, we risk reaching wrong conclusions.

Prais' method involves computing the regression of size in the final year on the size in the original year and vice versa.[9] These two regression coefficients are used to described the change in concentration. If both lines have a slope greater than one, concentration has increased, and vice versa. When the lines are on different sides of the 45°-line, the geometric mean of one coefficient and the inverted value of the other is decisive. According to Prais: "If this [the geometric mean] is greater than unity, concentration increases; if it is less, it decreases" (*Prais*, 1958, p. 272). This conclusion is based on statements to the effect that concentration increases when large firms show more than average growth, and vice versa.

Critical comment on this method is raised in *Adelman* (1959) and *Adelman & Preston* (1960). The latter study also applied the method to empirical data; the results obtained did not agree with results from earlier studies on changes in concentration. Adelman & Preston supposed that one reason for this discrepancy might be the entries and exits of firms.[10] The measure is applied in a study by *Gort* (1963) and is further commented on by *Horowitz* (1964).[11]

Joskow's method consists of analyzing the shifts in the ranks of firms. He proposes the correlation coefficient between ranks to summarize the measure of the shifts. Joskow claims that this measure could reveal mobility in industries which, according to other measures, seemed stable. *Hymer & Pashigian* (1962 a), however, argue that this measure has hardly any value, mainly because they suspect random variations to be larger in the lower classes. As an alternative to the measurement of changes in concentration they suggest use of an instability index, achieved by adding the absolute values of the changes in market share for every firm. The measure may be expressed:

$$I_t = \sum_{i=1}^{N_u} |P_{it} - P_{i,t-1}|, \qquad\qquad (5.16)$$

where

N_u = the union of the number of firms at time t and $t - 1$,

P_{it} = the market share of the ith firm at time t.

This measure takes entries as well as exits into account. It is also sensitive to two firms that are merely changing market shares, whereas measures of concentration are normally not affected by such changes. These two advantages are important enough to motivate the use of Hymer & Pashigian's index as a comparison to that of Adelman (see the subsection, "Measures of Mobility" below).

Empirical Section

Introduction

The techniques described in the preceding sections were applied to the data on the five samples of large firms (USA–SWED) for the period 1956–65. It is difficult to say whether or not this period covers different business conditions satisfactorily. This is especially true for areas consisting of several countries. The representation of difference business conditions with regard to USA and SWED seems good, in any case.

The classes have been chosen so as to make the upper class limits twice as large as the lower ones. The class ranges are expressed in absolute terms, and the firm sizes have been deflated to money value in the different years. This deflation was performed by means of the wholesale price index.[12]

A birth or death of a firm is regarded as having occurred when a firm passes the lower size limit of class 1. The usual method for dealing with mergers has been applied (see *Hart & Prais*, 1956; *Adelman*, 1958; and *Mansfield*,

1962a), i.e., the largest firm is said to have grown and the smallest to have disbanded.

Estimates of Transition Probabilities

The law of proportionate effect is assumed valid in the first estimations. The maximum likelihood method produces the results shown in table 5-1.
The following conclusions can be drawn from the table:

1. The probabilities of remaining in the same class are greatest
2. The probabilities of jumping more than one class are either zero or very slight
3. The probabilities of growth are higher than the probabilities of decline for all samples except USA*.

The results obtained are consistent with earlier findings. *Adelman* (1958, p. 900) as well as *Archer & McGuire* (1965, p. 238) also report the highest probabilities for remaining in the class. As for jumping more than one class, Adelman concludes that nature is regular and does not make jumps: "natura non facit saltum" (*Adelman*, 1958, p. 900). With respect to the probabilities of transition, Adelman obtained approximately the same probabilities for growth and decline, whereas Archer & McGuire found the transitions to the next higher class to be the more probable movement.[13]
Let us now drop the law of proportionate effect and estimate separate probabilities for the different classes. The results of this estimation are illustrated in tables 5-2 to 5-6.
The five tables lead to conclusions similar to those drawn from table 5-1. i.e., that the probability of remaining in a class is highest and that jumps greater than one class are very rare.
Earlier, we mentioned transition matrices as a device for examining the relevance of the law of proportionate effect. This can be done by comparing transition probabilities in different classes. The only observable tendency is that the growth probabilities for all samples are lower in the second class than in the first. The relatively small number of observations makes it difficult to draw any clear-cut conclusions, however. This is particularly true in the higher classes due to the skewness of the distributions.

Steady-State Distributions

The transition probabilities derived make it possible to compute the distributions in steady state. This was done with and without the law of pro-

Table 5–1
Estimates of Transition Probabilities when the Law of Proportionate Effect is Assumed Valid

Probability of jumping from class i to class j	$j = (i - 2)$	$j = (i - 1)$	$j = i$	$j = (i + 1)$	Σ
USA	–	.038	.892	.070	1.000
USA*	–	.067	.875	.058	1.000
EUR	.002	.066	.861	.071	1.000
SCAN	–	.025	.901	.074	1.000
SWED	–	.026	.894	.080	1.000

Note: The reason for the nonzero estimate of $p_{k, k-2}$ for EUR, when it is zero for the main group USA*, is differences in the development of the indices used.

Table 5–2
The Estimated Transition Matrix for USA

	0	1	2	3	4	5	6	Σ
0	.903	.095	.002					1.000
1	.042	.880	.078					1.000
2		.031	.903	.066				1.000
3			.033	.921	.046			1.000
4				.028	.889	.083		1.000
5						.889	.111	1.000

Table 5–3
The Estimated Transition Matrix for USA*

	0	1	2	3	4	Σ
0	.912	.085	.002	.001		1.000
1	.064	.875	.061			1.000
2		.069	.877	.054		1.000
3			.143	.786	.071	1.000
4				.063	.937	1.000

Table 5–4
The Estimated Transition Matrix for EUR

	0	1	2	3	4	5	Σ
0	.936	.062	.001	.001			1.000
1	.061	.865	.074				1.000
2		.073	.869	.058			1.000
3		.067	.067	.799	.067		1.000
4				.067	.866	.067	1.000

Table 5-5
The Estimated Transition Matrix for SCAN

	0	1	2	3	4	5	Σ
0	.924	.076					1.000
1	.024	.893	.083				1.000
2		.038	.892	.070			1.000
3				.945	.041	.014	1.000
4					.947	.053	1.000

Table 5-6
The Estimated Transition Matrix for SWED

	0	1	2	3	4	5	Σ
0	.956	.044					1.000
1	.025	.887	.088				1.000
2		.042	.881	.077			1.000
3				.938	.047	.015	1.000
4					.947	.053	1.000

portionate effect. The same number of classes are considered in both cases. The highest class taken into account is the last one, where p_{ii} can be estimated. This means that firms can leave the system from this class with no probability of return. This is interpreted to mean that we take only the present range of sizes into consideration.

This technique is used since we have no estimates of transition probabilities in classes beyond the present when the law of proportionate effect is not valid. When it is valid, we use it because most of the estimates of $p_{i,i+1}$ are larger than $p_{i,i-1}$. This fact implies that no steady-state solution is obtained for an unlimited number of cells (see the earlier section, "Steady-State Distributions"). The distributions derived are described by the parameter ρ in Simon & Bonini's model. The resulting values are shown in table 5-7.

A continuation of the process would yield an increase in the degree of concentration when measured with the Pareto coefficient in all areas. In most cases the process does not produce results consistent with the model of Simon & Bonini. If the distributions are plotted in double logarithmic scale, deviations from linearity can be observed. This may result from the assumption that firms can leave the population from the highest class. This kind of assumption is not made when using Simon & Bonini's model, but it is necessary here.

Since deviations from linearity exist, it might be appropriate to use a measure of concentration other than the Pareto coefficient. Thus values of the Gini coefficient were computed. The results are shown in table 5-8.

Table 5-7
Values of the Pareto Coefficient (ρ)

			Steady State	
Sample	1956	1965[a]	P	P*
USA	1.44	1.51	.71	.86
USA*	1.15	1.83	1.00	.94
EUR	1.15	1.69	.98	1.37
SCAN	1.89	1.52	.75	.71
SWED	1.78	1.34	.80	.68

Note: P_* = the law of proportionate effect is valid.
P^* = the law of proportionate effect is not valid.
[a]These figures differ to some extent from those presented in chapter 4 since another classification was used there.

Table 5-8
Values of the Gini Coefficient

			Steady State	
Sample	1956	1965	P	P*
USA	.409	.413	.446	.434
USA*	.397	.292	.382	.412
EUR	.397	.313	.383	.364
SCAN	.305	.347	.389	.387
SWED	.322	.359	.397	.390

Note: See note to table 5-7.

The table shows that the steady states imply an increasing degree of concentration relative to the state in 1965 for all areas. Thus distributions are not yet in steady state. In the samples USA, SCAN, and SWED the process is a continuation of the trend from 1956, but this is not the case for USA* and EUR. The values of the Gini coefficient in steady state are larger for all samples except USA* when the law of proportionate effect is assumed valid. The differences are very slight, however.

The magnitude of the values is similar for the five areas. Actually, a process of this type seems to imply a convergence of the concentration within the samples to Gini coefficient values of about 0.4.

Measures of Mobility

Transition probabilities may also be used to deduce the average time spent in a certain class and its standard deviation according to (5.11) and (5.12) above. This has been done for the four lowest classes and for the case where the law of proportionate effect is assumed valid. The values obtained are shown in table 5-9.

The table provides a basis for the following conclusions:

1. The average time spent in lower classes is about the same for the five samples of large firms. This also holds true when the law of proportionate effect is valid.
2. In all samples, the average time required to progress from the first to the fifth class is longer than the period of estimation.

The steady-state distributions derived also permit computation of Adelman's index according to (5.14). The values of this index have been compared with those obtained from Hymer & Pashigian's index. Data from 1956 and 1965 are used in computing the latter index. The results are shown in table 5-10.

Adelman's index yields values between 0.10 and 0.20 in the present study. Values of this magnitude were also reported in her own applications. No real pattern in the relation between values deduced with and without the law of proportionate effect may be discerned; the ranking is about the same (SCAN and USA* change places). The values obtained for Hymer & Pashigian's index rank the areas differently, but the index discriminates between samples of low (USA, SCAN, and SWED) and those of high mobility (USA* and EUR). The

Table 5-9
Mean and Standard Deviation of Time Spent in the Different Classes

Class	Sample	USA	USA*	EUR	SCAN	SWED
1.	$t^a_{i_a}$	8.33	8.00	7.41	9.35	8.85
	σ^a_i	7.81	7.48	6.89	8.81	8.36
2.	$t^a_{i_a}$	10.31	8.13	7.63	9.26	8.40
	σ^a_i	9.79	7.62	7.11	8.75	7.89
3.	$t^a_{i_a}$	12.66	4.67	4.98	18.18	16.13
	σ^a_i	12.15	4.14	4.72	17.67	15.63
4.	$t^a_{i_a}$	9.01	15.87	7.46	18.87	18.87
	σ^a_i	8.50	15.36	6.95	18.36	18.36
$\sum\limits_{i=1}^{4}$	t^a_i	40.31	36.67	27.48	55.66	52.25
P^1	$t^a_{i_a}$	9.26	8.00	7.19	10.10	9.43
	σ^a_i	8.75	7.48	6.67	9.58	8.92

[1] The law of proportionate effect is assumed valid.

Table 5-10

A Comparison between Two Measures of Mobility

Sample	USA	USA*	EUR	SCAN	SWED
Adelman (M_t) P	.138	.174	.193	.105	.166
P*	.091	.162	.229	.118	.129
Hymer & Pashigian					
(I_t)	.483	1.173	1.022	.405	.391

same kind of discrimination can also be achieved by using measures of concentration (see chapter 2). The three samples regarded as having a relatively low degree of mobility also have a relatively high degree of concentration.

Conclusions

The results presented in this chapter indicate that we can expect an increasing degree of concentration within the size range under study. But the conclusions of this study depend to a large extent upon the quality of our estimates. As a matter of fact, one may suspect some shortcomings in the estimates since the observation period and the populations are rather limited.

In any case, the general structure of the matrices derived seems to be in accordance with earlier results. That is to say, remaining in a class has the highest probability, and the only probabilities of any significance with respect to transitions are those expressing steps to adjacent classes. This is also as might be expected intuitively since absolute changes in size are relatively small compared to the initial sizes of firms. Average growth rates vary around 5 percent, which is to be compared to a step to the closest nonadjacent class involving a growth of 100 to 800 percent. Not even mergers can contribute to growth rates of this magnitude owing to the way mergers are treated (see the earlier "Empirical Section."). The maximum contribution of growth from a merger is 100 percent, which occurs when the merging firms are equal in size.

With respect to steady-state distributions, there are difficulties in comparing the actual and theoretical distributions. Differences do exist, although forces working toward the steady state can be observed.

One drawback in using the Markovian approach is that states other than the steady state are difficult to deal with, mainly because entries cannot be included.

Another disadvantage is that the analysis is limited to present sizes. This means that the real giants of the future are omitted from the analysis, although they are extremely important with respect to concentration. It should also be noted that steady-state solutions do not provide any information on the number of firms, but on proportions only.

A final deficiency is the constancy of the matrix, which implies an extrapolation of the trend. But, if we are trying to forecast future distributions, we are greatly interested in possible changes in the trend. A model which permits the matrix to change over time is needed to satisfy this interest. Such a model, based on the simulation approach, is presented in chapter 8. Before turning to such models, however, we will conclude our discussion of analytical models by considering two models using continuous size.

6 Models Using Continuous Size

In this chapter we will examine two analytical models that use continuous size. The models produce a distribution, frequently used in describing size distributions, known as the lognormal distribution, where the logarithms of values are normally distributed.

A great deal of interest has long been devoted to this distribution. An early study was carried out by *McAlister* (1879), and among the well-known researchers from the beginning of this century are Kapteyn and Van Uven.[1] These early studies mainly concerned phenomena in the physical sciences. Two studies by *Gibrat* (1930 and 1931) had a great impact on application of the lognormal distribution in the social sciences. These were followed by many studies with reference to economic phenomena, many of which deal with size distribution of firms. They include studies on British industries (*Hart & Prais*, 1956; *Hart*, 1957 and 1962; and *Singh & Whittington*, 1968), the largest industrial firms in France (*Morand*, 1967) and American firms (*Silberman*, 1964 and 1967).

The lognormal distribution has also been used to describe distributions of trade unions (*Hart & Brown*, 1957) and to analyze demand (see, e.g., *Holt et al.*, 1960; *Pessemier*, 1966; and *Magee*, 1967). A basic study in terms of methodology was undertaken by *Aitchison & Brown* (1957).

Models which generate lognormal distributions are discussed in the next section. We than proceed to sections on estimation and testing. Finally, the models are applied to empirical data.

The Models

Description

Gibrat presented a basic continuous model based on the following two assumptions: (1) the law of proportionate effect is valid, and (2) the number of firms is constant. Given these assumptions, it may be shown that logarithms of sizes are normally distributed.[2]

The change in size between two periods may be written as.

$$(S_t - S_{t-1})/S_{t-1} = \epsilon_t,$$ (6.1)

where

S_t = the size of a firm at time t

ϵ_t = a random variable.

Summing over T time periods we obtain:

$$\sum_{t=1}^{T} (S_t - S_{t-1})/S_{t-1} = \sum_{t=1}^{T} \epsilon_t. \tag{6.2}$$

If each period is assumed to be small,

$$\int_{S_O}^{S_T} 1/S \, dS = \sum_{t=1}^{T} \epsilon_t \tag{6.3}$$

or

$$\log S \Big|_{S_O}^{S_T} = \sum_{t=1}^{T} \epsilon_t \tag{6.4}$$

and

$$\log S_T - \log S_O = \sum_{t=1}^{T} \epsilon_t \tag{6.5}$$

or

$$\log S_T = \log S_O + \epsilon_1 + \epsilon_2 + \ldots + \epsilon_T \tag{6.6}$$

If T is large, $\log S_T$ has an asymptotically normal distribution on the basis of the additive form of the central limit theorem (see, e.g., *Lindgren*, 1962, p. 145). This means that sizes have an asymptotically lognormal distribution and that the frequency curve can be described by:

$$f(S) = (1/\sigma\sqrt{2\pi})e^{-(\log S - \mu)^2/2\sigma^2} \quad (S > 0) \tag{6.7}$$

where

μ = the mean of $\log(S)$,

σ = the standard deviation of $\log(S)$.

One of the consequences of Gibrat's model is that the variance increases steadily over time (see chapter 3, "Models Using Continuous Size"), which means an increasing degree of concentration over time. However, this is not the case in the modified model proposed by *Hart & Prais* (1956), which suggests that firm sizes tend to regress toward the mean as the firms strive to reach an optimum size.

According to Hart & Prais, the relation between the variance in two periods can be expressed:

$$\sigma_{t+1}^2 = \beta^2 \sigma_t^2 + \sigma_\epsilon^2, \tag{6.8}$$

where

σ_t^2 = variance of $\log(S)$ at time t

β = the regression coefficient

σ_ϵ^2 = the residual variance.

The correlation coefficient between $\log(S)$ in time t and ($t + 1$) is

$$\rho_{xy} = \sqrt{1 - \sigma_\epsilon^2 / \sigma_{t+1}^2}, \tag{6.9}$$

which may be written as follows:

$$\sigma_{t+1}^2 / \sigma_t^2 = \beta^2 / \rho_{xy}^2 \tag{6.10}$$

If $\beta^2 > \rho_{xy}^2$, the variance will increase, and vice versa. Thus Hart & Prais' model is also compatible with a decreasing variance. In contrast to Gibrat's model, the level of concentration remains undetermined.

The Assumptions

Concerning the first assumption—the law of proportionate effect—we may refer back to chapter 4 and the reasoning of Simon & Bonini. They argue that the law of proportionate effect can be accepted for firms having the same average costs. According to findings by *Bain* (1956), this holds true for large firms, which would then limit Simon & Bonini's analysis to this type of firm.

The basic model treated here deals with firms of all sizes. According to Simon & Bonini's argument, this means that the whole range of average cost curves is assumed to be horizontal.

Bain's investigations, however, point to falling average cost curves up to a certain size (see chapter 4). If this description is accepted, it would not be out of place to suspect deviations from the law of proportionate effect for the

non-giant firms. Since the latter face falling average cost curves, we may expect
their growth rates to be above the mean. This means that smaller firms exhibit
higher probabilities of moving upward along the continuum of sizes. For a
constant population the probable result is a lower number of small firms than
the model might otherwise lead us to expect. Furthermore, rejections may
occur when the law of proportionate effect is tested.

The modified model of Hart & Prais is also incompatible with Bain's
findings. Actually, they base their model on the assumption that cost curves
are U-shaped. This results in an underestimation of tendencies toward an
increased degree of concentration.

The validity of the second assumption, that the number of firms should
be constant, is also doubtful with reference to empirical data. In a discussion of
Hart & Prais' study, Champernowne concluded: "when many firms are being
born and dying, the theoretical leg for the log-normal distribution becomes
very shaky" (*Champernowne*, 1956, p. 1852). Thus we may expect the lognormal
distribution to be a less appropriate representation of a size distribution, when
the number of firms varies a great deal. The lognormal description might still be
relevant given a moderate change in the number of firms.[3]

Although the two assumptions may lack validity, they would tend to
counterbalance one another. Entries tend to reduce concentration, while, on
the other hand, the assumption that the law of proportionate effect is valid
for small sizes will probably lead to an overestimation of concentration.

Estimation Procedures

Lognormal distributions of firm sizes are very convenient for testing
purposes. Simple, well-known test statistics—such as Student's t—can be used.
Such tests require estimates of the distribution parameters, to which purpose
Aitchison & Brown (1957, p. 38) mention the following four methods:[4]

1.   The method of maximum likelihood estimates
2.   The method of moments
3.   The method of quantiles
4.   The graphical method

The first of these four methods, of course, is preferable when it can be computed. The estimates are obtained in the following way:

$$\mu_1^* = 1/N \sum_{i=1}^{N} \log S_i \qquad (6.11)$$

$$\sigma_1^* = \sqrt{1/N \sum_{i=1}^{N} (\log S_i - \mu_1^*)^2} \qquad (6.12)$$

where
μ_1^* and σ_1^* = the estimates of μ and σ from the maximum likelihood method.

The method of moments involves computing the parameters of the log-normal distribution from the first and second moments of the original distribution. Consequently, we can write:

$$\mu_2^* = 2 \log \left(\sum_{i=1}^{N} S_i/N \right) - 1/2 \log \left(\sum_{i=1}^{N} S_i^2/N \right) \qquad (6.13)$$

$$\sigma_2^* = \sqrt{\log \left(\sum_{i=1}^{N} S_i^2/N \right) - 2 \log \left(\sum_{i=1}^{N} S_i/N \right)}, \qquad (6.14)$$

where
μ_2^* and σ_2^* = the estimates of μ and σ from the method of moments.

The efficiency of this method declines rapidly with the size of σ. This means that it should not be used for σ-values above 0.5 (see *Aitchison & Brown*, 1957, p. 40).

The method of quantiles implies that two points in the distribution are chosen, such as $S_j < S_k$. The parameters are then estimated from:

$$S_j = e^{\mu_3^* + d_j \sigma_3^*} \qquad (6.15)$$

$$S_k = e^{\mu_3^* + d_k \sigma_3^*} \qquad (6.16)$$

where
μ_3^* and σ_3^* = the estimates of μ and σ from the method of quantiles.

The efficiency of this method is not affected by variance, but by the choice of d_k and d_j. The highest degree of efficiency is obtained when they are equal in size, i.e., when the quantiles are symmetrically located.[5] This is an interesting method of estimation, as some authors mention the geometric mean as a measure of concentration (see, e.g., *Niehans*, 1955). Thus, if we choose $d_k = 1$ and $d_j = -1$ and solve for the parameters we obtain:

$$\mu_3^* = \log \text{ (geometric mean of } S_k \text{ and } S_j) \tag{6.17}$$

$$\sigma_3^* = \log \text{ (geometric mean of } S_k \text{ and } 1/S_j) \tag{6.18}$$

The fourth method of estimation involves plotting some points (at least the deciles) of the cumulative distribution on lognormal probability paper and fitting a straight line to the plots. Percentiles of the distribution can then be read off. Using these observations, *Aitchison & Brown* (1957, p. 32) propose the following estimates:

$$\mu_4^* = \log S_{50} \tag{6.19}$$

$$\sigma_4^* = \log [0.5 (S_{50}/S_{16} + S_{84}/S_{50})], \tag{6.20}$$

where
$\mu_4^* =$ and $\sigma_4^* =$ estimates of μ and σ from the graphical method.

A simpler estimate of σ, suggested by *Pessemier* (1966, p. 173), is:

$$\sigma_4^* = \log S_{84} - \log S_{50} = \log (S_{84}/S_{50}). \tag{6.21}$$

The method of using lognormality paper may, of course, be criticized for its lack of precision, as the result depends to a large extent on the ability of the judges. The advantage in terms of simplicity seems to outweigh this criticism, however, especially with reference to large populations. Aitchison & Brown's method of overcoming this problem consisted of having various persons, not connected with their research, make the estimates (*ibid.*, p. 42).

In empirical applications for CAR and SHOE it was found that the method of quantiles was the best substitute for the maximum likelihood estimate, followed by the method of moments and the graphical method.[6] These results are consistent with those reported by *Aitchison & Brown* (1957, p. 44).

Test Procedures

In the previous section we mentioned the possibility of using simple, well-known test statistics for a lognormal distribution. Before these statistics can be applied, however, tests for lognormality have to be performed.[7] One possibility is to use a Kolmogorov & Smirnov test for goodness of fit.[8] In addition to such a test it also seems appropriate to use alternative methods as well. In earlier works we find the following suggestions:

1. Plotting the distribution on lognormal probability paper; linearity in this instance is a first indication of lognormality

2.   Check of the symmetry of the Lorenz curve
3.   Test of the correspondence between actual and theoretically derived concentration ratios
4.   Test of the third and fourth moments of the transformed distribution

The first two methods can be used as a first tentative test of lognormality. The principle of the first method is obvious, whereas the second is based on the fact that lognormal distributions produce symmetrical Lorenz curves (see *Aitchison & Brown*, 1954, p. 103, and *Hart & Prais*, 1956, p. 159).[9] However, a positive result in this test does not guarantee lognormality, since other distributions may also produce symmetrical Lorenz curves (see *Champernowne*, 1956, p. 182, and *Kendall*, 1956, p. 185).

An extension of the second method is to compare the actual values of the Gini coefficient with the theoretical value derived from an estimate of σ. The latter can be obtained from the following expression (see *Aitchison & Brown*, 1957, pp. 13 and 112):

$$R_\sigma = 2\,N(\sigma/\sqrt{2}\ |0,1) - 1, \tag{6.22}$$

where

R_σ = an estimate of the Gini coefficient from σ.

The implication of (6.22) is that an estimate of R can be obtained as follows:[10]

1.   Estimate σ
2.   Compute $z = \sigma/\sqrt{2}$ and calculate the corresponding value of the normal distribution function $\Phi(z)$
3.   The Gini coefficient is obtained from:

$$R = 2\,\Phi(z) - 1 \tag{6.23}$$

The third method is presented by *Silberman* (1967, p. 809 ff.).[11] It is based on the fact that theoretical values of concentration ratios can easily be estimated for a lognormal distribution with known parameters. The concentration ratios are derived from the first moment distribution having the mean $\mu + \sigma^2$ and standard deviation σ (see, e.g., *Aitchison & Brown*, 1954, p. 100). The estimation procedure runs as follows:

1.   Compute the percentile ratios $r_n = n/N$. For example, if the total number of firms is 40, and we want to compute CR_4, then $r_4 = 0.1$.
2.   Compute the number of standard deviations (z_n) in a normal distribution $N(0, 1)$ corresponding to r_n.

3.   Compute $z^* = (z_n - \sigma)$. [12]
4.   From a normal distribution table, derive the part of the first moment
     distribution corresponding to z^*. This is the estimate of CR_n looked for.

Estimates derived by this method are then used in Silberman's method of
testing in the following way:

1.   Actual concentration ratios are computed for, say, the 4th, 8th, 20th and
     50th largest firms.
2.   Theoretical concentration ratios are computed assuming lognormality.
3.   The standard deviations of the concentration ratios are computed by
     means of simulation.
4.   The actual value for the four firms is first tested against the theoretical
     value. Should the difference be significant, the lognormal hypothesis is
     rejected. Otherwise, we go on to the next concentration ratio and perform
     the same procedure.

Silberman argues for starting with four firms and then proceeding in the
distribution since he regards the upper tail as the most important and expects
the differences between the lognormal distribution and other skew distributions
to be most pronounced in this range (*Silberman*, 1967, p. 819). Silberman's
concentration of attention on the tail of the distribution was subsequently
supported by *Quandt* (1968), although Quandt is critical of the method of
testing. He does not agree with Silberman's assumption that concentration ratios
have a normal distribution and thus suspects that the method proposed is biased
against the rejection of lognormality. In a reply to Quandt, *Silberman* (1968)
refers to tests with the Kolmogorov & Smirnov test, which he performed as a
check. Application of this test produced the same result as the method originally
used.

The fourth method mentioned above was first proposed by *Fisher* (1936)
and was used for size distributions of firms by *Hart & Prais* (1956). Use of this
method involves calculating the third and fourth moments of the transformed
distribution. Sampling errors are also estimated. The third moment involves
checking the symmetry, and the fourth, tests for kurtosis (convexity). In a
normal distribution these two moments have an expected value of zero. But
this property also characterizes other distributions, and consequently the test
does not guarantee normality. Thus it will not be used below.

Empirical Section

The Basic Model

In this section the lognormal distribution is applied to the distributions of
Swedish establishments in 1952 and 1966. Some other distributions, among

them the socialist samples, are also examined briefly. The testing methods used
are those described in the preceding section. The four distributions are first
shown in lognormal probability scale in figures 6-1 and 6-2. As is shown here,
linearity is most pronounced for SHOE. The population CAR in 1952 and 1966,
on the other hand, exhibit deviations from linearity. These are larger for the
distribution in 1966 than in 1952.

Drawing the Lorenz curves for the four distributions, figure 6-3 is obtained.
There are no strong indications of asymmetry in these curves, and we thus
proceed in testing the four populations. In so doing we need estimates of
distribution parameters. Using the method of maximum likelihood, we obtain
the estimates shown in table 6-1.

The estimates shown in table 6-1 have been handled according to the
procedure described in the last section in order to obtain theoretical values of
the Gini coefficient. The estimates obtained are shown in table 6-2, where they
are compared with values obtained using Yntema's formula.[13] The table shows
that all the estimates obtained from the standard deviation of logarithms are
smaller than the values from Yntema's formula. The differences are only slight,
however, and exceed 0.1 in only one case (CAR, 1966). Our results imply that

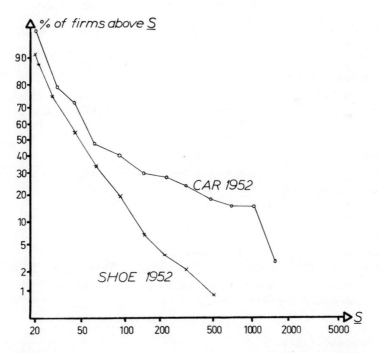

Figure 6-1. Size Distributions of Establishments in CAR and SHOE:
1952

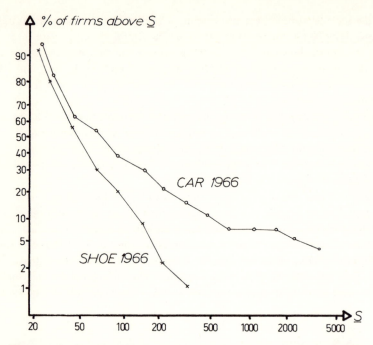

Figure 6-2. Size Distributions of Establishments in CAR and SHOE: 1966

concentration is underestimated, which in turn can be interpreted to result from an underestimation of σ.

The values of concentration ratios obtained are displayed in table 6-3.

The differences found for CAR are consistent with *Silberman*'s (1967, p. 819) view that the uppermost part of the tail is the most critical area in the distribution. CR_4 deviates the most for CAR, the differences diminishing thereafter. This does not hold true for SHOE, which might imply that this population deviates slightly from a lognormal distribution. We should also note that the differences for this population are greater in 1966 than in 1952 for CR_4 and CR_8

As a final check a Kolmogorov & Smirnov test on a 1 percent level of risk was applied, producing the results shown in table 6-4.

Out of the four values, one is significant, namely, SHOE in 1952. There does not seem to be any reason to reject the lognormal hypothesis for the other three populations on the basis of this test.

The Model of Hart & Prais

There are no apparent tests for Hart & Prais' model, but we can compute the parameters of the model, i.e., the coefficient of regression and correlation

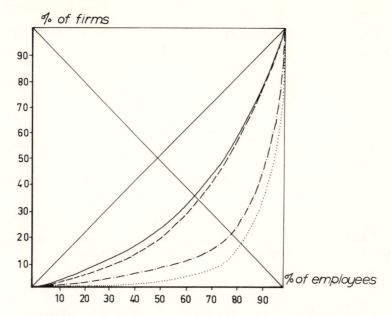

Figure 6–3. Lorenz Curves for the Samples CAR and SHOE: 1952 and 1966

The curves represent SHOE 1966 _____ ,
SHOE 1952 _____ , CAR 1952 _.._.._.. .
and CAR 1966 . . . .

Table 6–1
Estimates of the Parameters in the Lognormal Distributions (Natural Logarithms Are Used)

Sample	CAR		SHOE	
	1952	*1966*	*1952*	*1966*
μ^*	4.595	4.466	3.927	3.933
σ^*	1.289	1.340	.716	.654
Number of firms	37	80	131	91

for the period 1952–1966. The values obtained are shown in table 6- 5. The figures in the table illustrate deviations from unity for the regression coefficient in both samples. The figure is above this value in CAR and below it in SHOE. Concentration can therefore be expected to increase in CAR, since the coefficient of correlation is always less or equal to unity. A decrease in concentration in SHOE is indicated since b is less than r for this population.

Table 6-2
Values of the Gini Coefficient

Sample	CAR		SHOE	
Value	1952	1966	1952	1966
Estimate from σ^* (R_σ)	.639	.657	.383	.365
Estimate from Yntema's formula	.709	.801	.438	.391
Difference	−.070	−.144	−.055	−.026

Note: The values of R_σ are taken from table A1 in Aitchison & Brown (1957, appendix). Note that the table expresses the standard deviation in natural logarithms.

Table 6-3
Theoretical and Empirical Values of Concentration Ratios

Number of firms	4	8	16	32	64
CAR 1952					
Theoretical value	.520	.693	.868	.992	
Empirical value	.638	.808	.910	.988	
Difference	−.118	−.115	−.042	−.004	
CAR 1966					
Theoretical value	.380	.513	.691	.862	.999
Empirical value	.636	.769	.854	.930	.987
Difference	−.256	−.256	−.163	−.068	.012
SHOE 1952					
Theoretical value	.125	.203	.327	.509	.754
Empirical value	.178	.271	.390	.568	.777
Difference	−.053	−.068	−.063	−.059	−.023
SHOE 1966					
Theoretical value	.023	.043	.392	.608	.882
Empirical value	.177	.286	.439	.641	.880
Difference	−.154	−.243	−.047	−.033	.002

Table 6-4
Values Obtained in the Test for Goodness of Fit

	CAR		SHOE	
Sample	1952	1966	1952	1966
Largest deviation	8.162	8.104	8.094*	8.083
Critical value	8.169	8.115	8.090	8.108
Number of observations	37	80	131	91

Table 6-5
Parameters for CAR and SHOE in the Regression Model of Hart & Prais

Sample	CAR	SHOE
Regression coefficient (b)	1.125	.630
with standard deviation (s_b)	.083	.085
Correlation coefficient (r)	.947	.696
b^2/r^2	1.440	.819

The expression $(b/r)^2$ provides information about the expected relation between the variance of the distributions in 1966 and 1980 since the model assumes that the relation is the same for both the fifteen future and fifteen past years. Thus the standard deviation in CAR will increase to the value 1.61.[14] The standard deviation in SHOE can be expected to take a value of about 0.59.

The changes in concentration indicated by the regression model can also be compared with the values of the index proposed by Hymer & Pashigian.[15] This index yields the value 0.899 and 0.927 for CAR and SHOE respectively. According to this method of measuring, there were more changes in the positions of firms in SHOE than in CAR, but the regression model indicated a greater change in concentration in CAR than in SHOE. This might be interpreted to mean that firms in SHOE changed places to a larger extent without influencing concentration. Thus the results are probably not as contradictory as they might seem at first glance.

Testing the Assumptions

In testing the two assumptions of the model the techniques used in chapter 4 were used here with respect to the law of proportionate effect. The values obtained from the computation of the regression coefficient are shown in table 6-6. The table shows one significant value, namely, that for SHOE in 1952-53. The test value in 1965-66 for the same population is not significant at the chosen level (1 percent.). The probabilities of transitions that express upward or downward movements to an adjacent class are shown in table 6-7. The number of observations is small, and the only conclusion we can draw is that a visual examination does not indicate any systematic pattern in the probabilities. Thus, in conclusion, the tests of the law of proportionate effect have not indicated rejection of this assumption for CAR. But it does appear to be a less relevant assumption for SHOE, at least in the first year.

As for the second assumption about the constancy of the population, figure 6-4 shows the development over time. The diagram reveals sharp changes in the number of firms for both CAR and SHOE. The former population

Table 6-6
Values Obtained from Testing the Law of Proportionate Effect

	CAR		SHOE	
Sample	*1952–53*	*1965–66*	*1952–53*	*1965–66*
Coefficient of regression (b)	1.005	.975	.817	.921
Standard deviation (s_b)	.020	.016	.034	.033
Degrees of freedom	35	72	129	87
t-value	.244	−1.460	−5.446[*]	−2.374
Critical value	2.73	2.65	2.61	2.64

Table 6-7
Probabilities of Transitions Up or Down One Class in CAR and SHOE

	CAR		SHOE	
Class	*Down*	*Up*	*Down*	*UP*
1	.142	.110	.112	.083
2	.052	.105	.119	.040
3	.135	.052	.084	.071
4	.036	.107	.229	−
5	.094	.094	.061	−
6	.071	.215	.250	−
7	.071	.108	.333	−
8	−	.037	−	−

Note: Estimation period: 1952–66.

underwent a rapid change in the amount of units—about 75 percent—during the sixties. The trend for SHOE is exactly the opposite. From 1957 to 1966 this distribution decreased from 136 to 91 firms, which amounts to a decline of 35 percent.

The number of firms in both industries during the first years of the period is fairly stable. This means that we should expect to find the lognormal distribution more appropriate for the distributions in 1952 than in 1966. This expectation is met for the population that changed the most, i.e., CAR. In the diagrams, the deviations from linearity were more pronounced for this population in 1966 than in 1952. We also observed increased differences between theoretical and actual concentration ratios for CAR in 1966 (see the section, "The Basic Model," above). These conclusions are consistent with a hypothesis proposed by *Wedervang* (1965, p. 85), who states that contracting sectors will better adhere to a lognormal distribution than expanding sectors.

It seems quite natural to accept the method used by *Hart & Prais* (1956, p. 165 ff.) for observing changes in the population. They computed the parameters

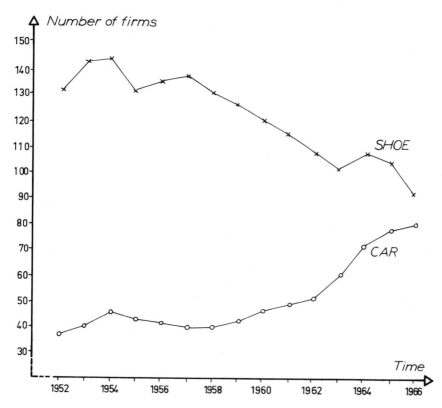

Figure 6-4. Numbers of Firms over Time in the Two Populations CAR and SHOE

for firms that survive, exit and enter. The results of calculations for CAR and SHOE are illustrated in table 6-8. Our results are similar to those of Hart & Prais. Entering and exiting firms exhibit lower values for the lognormal parameters than the total group of firms. The change in the population SHOE is also noteworthy. More than half of the population in 1952 has gone out of business in 1966. Entries of about thirty did reduce the net effect of this high number of exits.

Other Populations

One might suspect that all the above conclusions are specific for the two populations CAR and SHOE. In order to check this aspect all Swedish industries in 1966 (= the final year used above) and the socialist samples were used. Since only grouped data were readily available, the mentioned distributions are only

Table 6-8

Changes in the Number of Firms and the Lognormal Parameters in the Two Establishment Distributions

Population	CAR			SHOE		
Subgroup	N	μ	σ	N	μ	σ
Total in 1952	37	4.595	1.289	131	3.927	.716
Exiting in 1952	14	4.329	1.036	71	3.773	.642
Surviving in 1952	23	4.763	1.369	60	4.161	.735
Surviving in 1966	23	5.150	1.627	60	4.174	.665
Entering in 1966	57	4.270	1.080	31	3.591	.386
Total in 1966	80	4.466	1.340	91	3.933	.654

Notes: We might have overestimated entries and exits. This can happen when the names of firms are changed. The identification numbers have also been changed by the census people in 1959–60 and 1964.

The exiting and entering firms changed status some time during the period, but the measurement was made in 1966.

Natural logarithms are used.

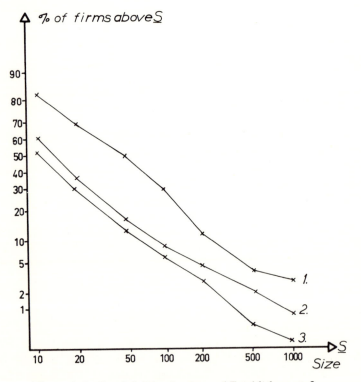

Figure 6–5. Swedish Distributions of Establishments I
(Lognormal Probability Scale)

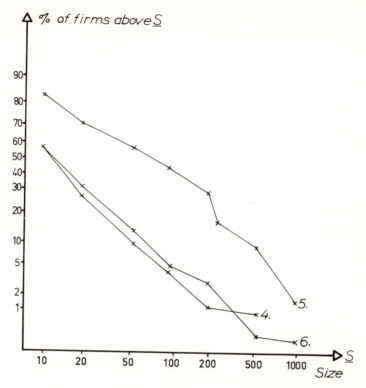

Figure 6-6. Swedish Distributions of Establishments II
(Lognormal Probability Scale)

plotted in lognormal probability scale. The results for the distributions of all
Swedish industries in 1966 are shown in figures 6-5 through 6-8.[16]

The distributions show fairly good linearity, with the exception of industry
8—the beverage and tobacco industry. One interpretation of the curved line
might be that there are "too many" small firms in this industry. This, in turn,
could be a result of the shape of the cost curves.[17]

When the data from the socialist countries are plotted in lognormal
probability scale, the results shown in figure 6-9 are obtained.[18] As shown
in this figure, a linear description seems equally reasonable for the socialist
countries as for the nonsocialist countries. There is no indication that the
straightness is less pronounced for the former samples than the latter, which
would imply that there is no less reason to consider lognormality as a description
for socialist size distributions than for nonsocialist ones.

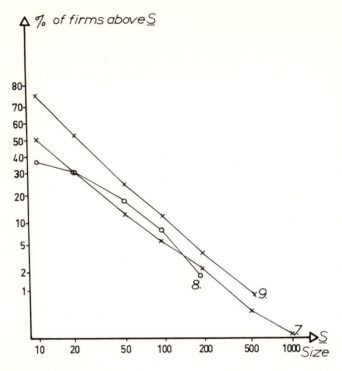

Figure 6-7. Swedish Distributions of Establishments III (Lognormal Probability Scale)

Conclusions

Silberman found the lognormal distribution to be an appropriate description for only 50 percent of American industries and therefore concluded that the lognormal distribution is not a good general method for describing size distributions (*Silberman*, 1967, p. 809).

An alternative conclusion is that the lognormal distribution is most appropriate for industries exhibiting moderate change in number of firms. This explanation also seems important in connection with a second conclusion suggested by Silberman, namely, that the lognormal distribution appears to be more representative of company distributions than establishment distributions (*ibid.*, p. 829). No applications for company distributions were made here, but we did obtain fairly good results for establishments. Similar results are also reported in a study by *Wedervang* (1965). A possible explanation of the differences in results is that American industries have undergone a greater change in the number of establishments.

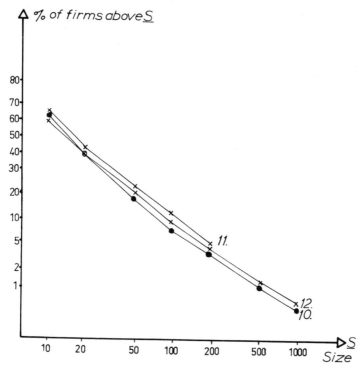

Figure 6-8. Swedish Distributions of Establishments IV (Lognormal Probability Scale)

Another interesting point made by *Silberman* (1967, p. 827) is that the lognormal hypothesis is rejected more frequently for industries comprising fewer than 10-200 business units. *Wedervang* (1965, p. 85) came to a similar conclusion, finding the lognormal distribution more appropriate for complete distributions than for truncated ones. None of our four samples fulfill these requirements, since they are all truncated samples with fewer than 200 business units. We have also found some indications of deviations from lognormality in our tests. Thus, our findings seem to be somewhat consistent with those of Silberman and Wedervang.

In the preceding section we also found the lognormal distribution to be a reasonable description of size distributions in socialist countries. Concentration may therefore be suspected of being something more than a peculiarly capitalistic phenomenon. Consistency between Gibrat's basic model and the results for the socialist countries is perhaps not as strange as it might first appear. If a size distribution of firms is lognormal, the law of proportionate effect must have been working on a relatively constant population. In the case of the

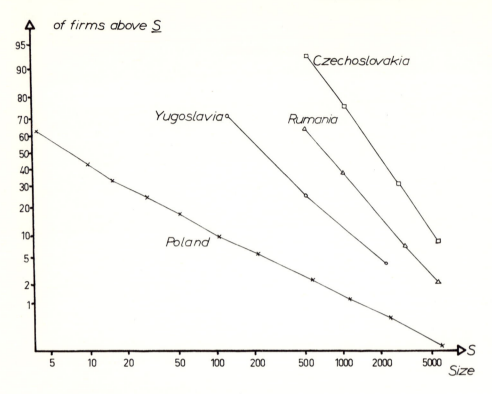

Figure 6-9. Size Distributions of Industry as a Whole in Four Socialist Countries (Lognormal Probability Scale)

socialist countries this might be because the economic administrators have been able to control entries and growth probabilities of firms.[19]

We have now considered analytical models using discrete as well as continuous size. The knowledge gathered in chapters 3 to 6 will now be used in developing new models by employing the simulation approach.

**Part 3
Simulation Models**

7 Simulation

Up until now we have considered analytical models. Although they can provide useful information in many instances, they have limitations connected with their assumptions and with difficulties in forecasting future market structures. As a consequence, a more flexible approach is in some cases desirable. A useful technique in this context is simulation, an approach increasingly applied during the last decade.

Part 3 will deal with models employing simulation. The present chapter goes into the technique and some of its characteristics. Moreover, earlier works using simulation related to size distributions of firms will be discussed. Then chapters 8 and 9 present two new simulation models, based on the models presented in Part 2: chapter 8 discusses a simulation model using discrete size, whereas chapter 9 deals with a model using continuous size.

The Simulation Technique

The simulation approach implies that experiments are performed not on the real system but on a model of this system. In many instances this technique is very useful, as manipulations of the real system are unfeasible due to limited funds or other restrictions.

The purpose of simulation is then, as Shubik expresses it, to study the operation of the model, "and, from it, properties concerning the behavior of the actual system or its subsystems can be inferred" (*Shubik,* 1960, p. 909).

Simulation can take many forms, such as experiments with scale models in designing ships and airplanes, training of air pilots in certain devices, and so forth. We will not, however, be concentrating on simulation in these terms here. Instead, the focus will be on an aspect of the concept which has arisen during the last two decades: computer simulation. This implies the construction of a model that is "a logical-mathematical representation of a concept, system, or operation programmed for solution on a high-speed electronic computer" (*Martin,* 1968, p. 5).

The significance of the computer in this context arises from its ability to run and manipulate complicated models over long periods of time with many iterations. These properties have been very important for the use of the simulation technique in many areas of the social sciences. We should also mention the flexibility offered by the simulation technique, implying that empirical and theoretical research can be combined to a higher degree.[1]

Simulation models on many different levels of description have been applied to subjects related to economics.[2] Thus *Adelman* (1963), *Duesenberry et al.* (1960) and *Klein* (1950) are examples of studies focusing on a whole economy; *Balderston & Hoggatt* (1962) and *Cohen* (1960) constitute examples of works studying different industries, whereas the firm is the research object of *Bonini* (1963), *Cyert & March* (1963), *Forrester* (1961) and *Hoggatt* (1957).

In building simulation models the purpose is not, of course, to obtain a one-to-one relationship. Instead, simulation models—like most models—stress some important features and exclude others. For example, in building test models of ships and aircraft the shape is often stressed, while the size is reduced. Similarly, when a system is imitated in a computer program, only the most relevant characteristics are included.

An appropriate question in model-building is, of course, how the relevant characteristics should be chosen. This question is particularly important with respect to computer models, since it is technically possible to include almost anything in a simulation model. The main limitation is that increased complexity of a model limits its comprehensibility and that of its output. Thus one must make a tradeoff between *realism* and *simplicity*. [3]

Selection of the model components should, of course, be made with regard to the significance ascribed to the various features. Thus, theoretically, if we start with a simple model and move toward a more complex one, the components added should be of progressively decreasing significance. Conversely, if we start with a complex model and move toward a simple one, the components dropped should have an increasing degree of significance.[4]

It is quite difficult to make such additions or reductions in a model, since in many cases the purpose of the research is to find the important components. As a consequence, validation of simulation models becomes extremely important. Basically, validation implies tests of how well the output of a model corresponds to that of the real system.

It need hardly be mentioned that the task of validation involves several difficulties. Amstutz summarizes the nature of these difficulties in the following way:

> Tests of validity are concerned with 'truth'. While reliability may be assessed using normal statistical techniques, there are no objective measures of truth. . . . In the absence of objective measures, the researcher must turn to a subjective evaluation of the consistency of the model's performance with theory and prior practice. (*Amstutz*, 1970, p. 378)

Naylor & Vernon suggest two test criteria for the validation of simulation models:

> First, how well do the simulated values of the endogenous variables compare with known historical data, if historical data are available? Second, how accurate are the simulation model's predictions of the real system in future time periods? (*Naylor & Vernon, 1969, p. 341*)

Many authors suggest that the second type of validation is to be preferred to the first, since simulation permits manipulation of parameters in order to obtain the "right answer." Thus, with regard to this method, Naylor et al. write:

> It is our position that the ultimate test of a computer simulation model is the degree of accuracy with which the model predicts the behavior of the actual system (which is being simulated) in the future. (*Naylor et al., 1966, p. 318*)

The drawback of this method is, of course, that a certain amount of time has to elapse before the validation can be performed.

As for the goodness of fit, several statistical techniques can be used, such as nonparametric tests (e.g., the Kolmogorov-Smirnov test), regression analysis, and factor analysis (see *Cohen & Cyert*, 1961, p. 120). In some instances methods like Student's t may be used to determine whether actual parameter values could have been drawn from the distribution of values obtained in the simulation runs. We will use this method in the following two chapters.

Two Earlier Simulation Models Related to Industrial Structure

Two earlier simulation models related to industrial structure ought to be mentioned. They have been developed by *Balderston & Hoggatt* (1962) and *Ijiri & Simon* (1964).

The first two authors simulated the lumber market of the United States, considering three levels: producers, wholesalers, and retailers. In their model the communication link between producers and retailers always goes via the wholesalers, who work as clearinghouses for different bids. The model was run with two kinds of preference formation processes in the three markets. Thus the firms made their preference ordering using experience in some runs and random choice in others. The flow chart in figure 7–1 summarizes the model.

The authors found that the different preference rules used and the cost of information were decisive for the shape of the size distributions (*Balderston*

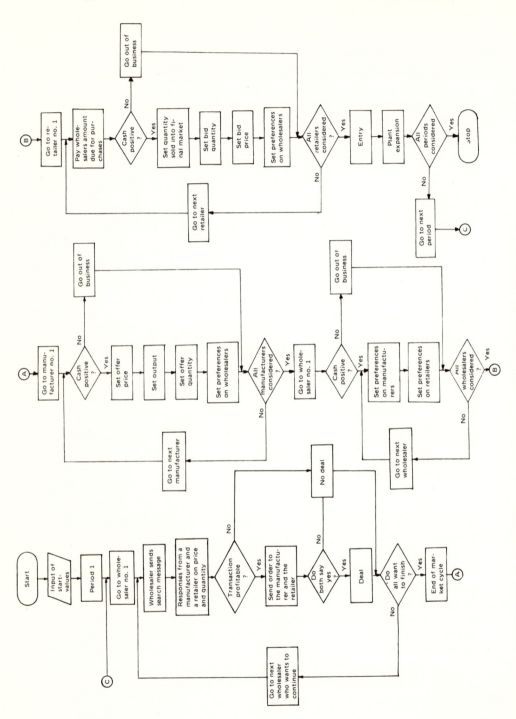

Figure 7-1. A Summary of Balderston & Hoggatt's Model

& *Hoggatt,* 1962, p. 37). Starting with equal size distributions, their results showed that the inequality of the size distribution of retailers and wholesalers increased ceteris paribus as (1) message costs increased, or (2) experience was used for the preference ordering (*ibid.,* p. 122 ff.). Of these two groups of firms, the distributions of those containing wholesalers are more skewed than those containing retailers under the same conditions, since the latter work under less market pressure.

With respect to the distribution of supplier firms, Balderston & Hoggatt argue that increasing message costs will reduce the wholesalers' chances of influencing the behavior of suppliers. Consequently, they hypothesize: "we should expect to find a more unequal size distribution of suppliers at low message costs than high ones" (*ibid.,* p. 125). Their results do not agree with this hypothesis, however, and they conclude that these firms "present a more complex pattern of experience, . . . but this experience does, in every case, show increasing differentiation of firm size from the initial time period onward" (*ibid.,* p. 125). *Balderston & Hoggatt* (1962, p. 146) also relate their results to stochastic theories and conclude in this context that their findings are a quantitative demonstration supporting these theories.

The model used by Balderston & Hoggatt has been developed by *Preston & Collins* (1966), who, with respect to size distributions computed "the coefficient of variation of firm sizes for wholesalers and suppliers for each period and the mean value and trend for this statistic over each run" (*ibid.,* p. 77). Their results indicate large variations in the dispersion among runs. However, the trends are positive in a majority of the cases, which is in accordance with the results of Balderston & Hoggatt.

A simulation model more closely related to the models discussed in the previous three chapters is presented by *Ijiri & Simon* (1964). Their model assumes a weaker form of the law of proportionate effect, which makes it possible to introduce growth sustenance into the model. This weaker variant postulates that "the expected percentage change in size of the *totality of firms in each size stratum* is independent of stratum" (*ibid.,* p. 79). The model adds one size unit at a time to the size units of the industry in question. The allocation of these additional size units is then determined in either one or two steps. First, the model determines whether the unit shall be allocated to an old or a new firm. Second, in the case of an old firm, the recipient is chosen with respect to the existing growth potentials of individual firms.

These growth potentials are determined by serial correlation, giving more weight to recent growth than to earlier growth. The unit is then allocated by drawing a

> random number r_2 from a rectangular distribution between 0 and $w(t-1)$, and assign the unit to the kth firm where k is the largest integer which satisfies

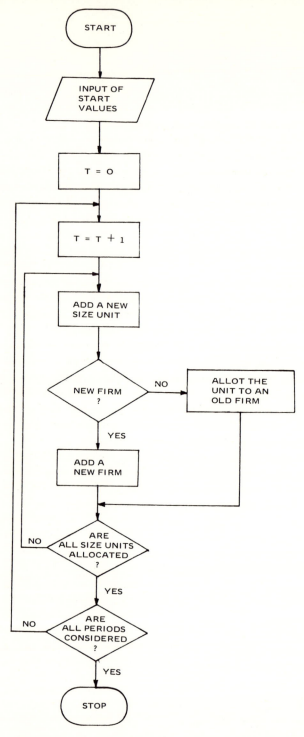

Figure 7-2. A Summary of Ijiri & Simon's Model

$$\Sigma \, w_j(t - 1) \leqslant r_2 \, \ldots \, (ibid., \text{ p. 83}),$$

where $w(t - 1)$ = the joint growth potential of all firms in period $(t - 1)$

$\qquad w_j(t - 1)$ = the growth potential of the jth firm in period $(t - 1)$.

This model can be summarized as in figure 7-2.

Ijiri & Simon found that their model produced size distributions of the skew type found empirically for U.S. industries. Similar results were found by *Imai* (1966), applying the model to Japanese manufacturing industries.

Conclusions

In this chapter we have briefly described the simulation technique, which, thanks to the computer, has been increasingly used since the early 1950s. We have noted the advantages of the approach as well as some of the problems involved in its use. With respect to the latter, we have particularly focused on problems of representation and validation. Moreover, two earlier models related to industrial structure have been described. One of these is also directly related to certain earlier stochastic models. In the following two chapters we will proceed in our discussion of such model-making by presenting, testing, and applying two new models. The main advantages of these models in comparison to those presented in Part 2 are that (1) the assumptions can be discarded if desired, and (2) information on size distributions in a specific time period can more easily be acquired. We will start with a model using discrete size and then proceed to one using continuous size.

8

A Simulation Model Using
Discrete Size

As was indicated in the previous chapter, the simulation technique has many advantages which make it useful in analyzing size distributions of firms. One advantage with respect to the models using discrete size is that states other than steady states can be considered. Hence, predictions may be made about industrial structure at specific points of time. However, if other states than steady states are analyzed, we cannot incorporate the entry process in the matrix, as was done in chapter 5.[1] Instead, the model has to contain two processes, one determining changes in size, another determining entry. They interact in the manner summarized in figure 8-1.

The following section deals with the two processes of the model, whereas the next discusses estimation procedures. Then results from work on validation and forecasting are presented in the following two sections.[2]

The Processes of the Model

As in the analytical models, the sizes of firms and exits out of the population are determined by a transition matrix. Entries, however, are governed by a mechanism of their own. This means that all rows except the zeroth row are included in the matrix.[3]

The revision of the entry mechanism is the basic difference between the present model and analytical models discussed earlier. However, other changes can also be made as they appear relevant. As was noted in chapter 7, a simulation model can include almost everything, a circumstance which gives rise to the problem of choosing the relevant modifications. Here we will start with a simple model and move toward a more complex one. Thus features of decreasing significance will be added as the model is made more complex, i.e., the most important modifications should be made first. (See chapter 7, "The Simulation Technique".)

Here it is appropriate to recall the basic assumptions of the analytical models, as they are the simplifying components of these models. Thus the transition toward a more complex model actually implies the relaxation of the assumptions completely or in part. The assumptions of the analytical models can be summarized as follows: (1) all firms are influenced by the same mechanism, and (2) the mechanisms are constant over time.

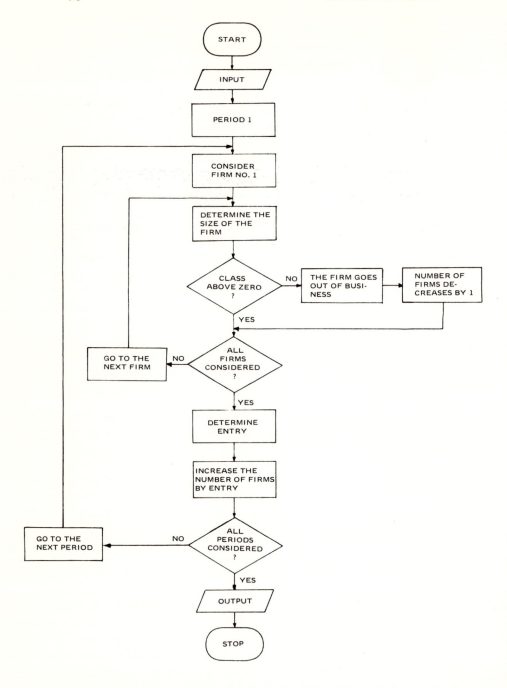

Figure 8-1. A Summary of the Model Using Discrete Size

Regarding the first assumption, the omission of the law of proportionate effect implies a partial withdrawal. However, no consideration is yet taken of the earlier performance of the firm. Such relations should be considered if we care to retreat further from the first assumption. We will here suggest three possible modifications with respect to the mechanism of changes in size. These are that the model takes into account: (1) past patterns of changes in size, (2) the time spent in a class, and (3) special properties exhibited by the firms. In the first case, we expect a firm that has experienced above-average growth one year to have above-average growth probabilities the following year. This is in accordance with studies carried out by *Ijiri & Simon* (1964).[4]

In the second case, newcomers in a class are assumed to have a high probability of decline and a low probability of growth. It is assumed that this situation changes as time goes by. Thus, over time, the probabilities of decline and growth decrease and increase, respectively. These circumstances have been verified for the lowest class (see *Lindgren*, 1949; *Wedervang*, 1965; and *Morand*, 1967). In other words, death risk declines with the age of the firm. It seems reasonable, however, to assume that the same situation prevails in other classes. Every class can then be assumed to be the lowest in a population where the classes below are excluded. In this context, the death of a firm means a step down to the next lower class. In the same way, entry corresponds to a step up to the next higher class.

Using this approach, the probabilities are estimated with reference to age groups of firms. It might be possible, for example, to find some regression relation between probabilities and time. When using this method we cannot completely omit the assumption that the law of proportionate effect is valid. Otherwise, the number of observations underlying the estimations would be too few. However, the interpretation of the law differs from the one generally used in that the mean probability of a certain growth rate for firms within a class is independent of size class, i.e., a firm that has been in class 2 for five years has the same probability of growth as one remaining in class 3 for the same length of time.

The third case generates the question of identification of the important factors with respect to changes in size. Examples of such factors are:

1. Branch characteristics, price trends, etc.
2. Propensity to do research
3. Propensity to make investments
4. Productivity/employee
5. Competence of management

In order to incorporate such characteristics in the model, they must be used in the classification of the firms. Different probabilities of change may then be estimated for firms showing high, normal, and low values in this classification.

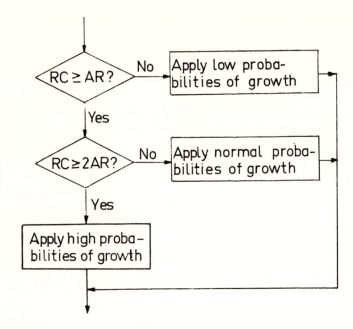

Figure 8-2. Mechanism Covering the Propensity to Do Research

Explanations: RC = costs for research and development in % of
turnover for the firm under study.

AR = a constant determined on the basis of the research
activities in the population under study.

Figure 8-2 illustrates an example of this type of modification where the propensity to do research has been included in the model.

Turning to the second assumption—the constancy of the mechanism over time—critics have frequently suggested a relation to the business cycle. Hart, for example, states

> we should expect [the probabilities] to vary between slump and boom on the one hand, when the mobility of firms between size-classes can be expected to be high, and periods of war and stagnation, on the other hand, when we should expect size-mobility to be generally lower. (*Hart*, 1962, p. 37)

Actually, this criticism can be countered if different business conditions are covered satisfactorily. One such alternative would be to estimate different probabilities for slump years and boom years.[5] We can then assume some sort of regularity in the cycle, e.g., that two years of slump are followed by two years of boom, and so forth. In the simulation, we can employ some random mechanism to determine the business condition during the starting year. The two matrices are then used alternately in the regular manner described above. Naturally, disturbances in regularity may also be introduced by some random mechanism.

The entry mechanism of the model generates an inflow of firms every year. This is accomplished in two steps: first, the number of entrants is determined, and then the entering firms are assigned to size classes.

Of the two assumptions mentioned above, that regarding the constancy of the mechanisms over time pertains to entry. Thus it might, for example, be appropriate to include cyclical variations in the entry mechanism as a result of changing business conditions.

It has to be stressed again—as we did in chapter 7—that all modifications will increase the complexity of the model, which gives rise to difficulties in analyzing the output. Moreover, the modifications will also impair the accuracy of the probability estimates, since they will reduce the number of underlying observations. Such problems and suggestions for estimation procedures are dealt with in the following section.

Estimation Procedures

The procedures suggested in chapter 5 in equations 5.1–5.4 may be used for estimations of changes in size. In introducing some new features, caution is necessary so that the estimates are not based on too few observations. The linear programming approach is one way of dealing with this problem (see chapter 5, "Estimating Transition Probabilities").

Subjective probability estimates, where experts are asked to assess the probabilities of certain events, are a method of estimation to be discussed further in the following chapter. Although they could be used here, they are more difficult to apply and thus will not be further discussed in this chapter.[6]

The entry mechanism is governed by a probability distribution for certain entries. The basic method for estimation in this context is to proceed as follows:

1. The number of firms during the observation period is recorded.
2. The estimate of the probability that m firms enter is expressed as:

$$p(m) = \sum_{t=1}^{T} b_t/T, \tag{8.1}$$

where $b_t = 1$, if m firms enter during period t, otherwise $= 0$
T = period of estimation.

3. The probability that one of the entering firms will go into the jth class is then expressed as:

$$p_j = \sum_{t=1}^{T} a_{0jt} / \sum_{j=1}^{nc} \sum_{t=1}^{T} a_{0jt},\tag{8.2}$$

where a_{0jt} = the number of new firms entering the jth class in period t
nc = the number of classes.

This method can be used when the observation period is long enough to allow a sufficient number of observations. However, in practice, the entry probabilities usually have to be estimated according to the following short-cut method:

1. Observe the number of entering firms during the observation period.
2. Determine the range of entries, i.e., the largest number of entries minus the smallest number of entries.
3. Regard all events within the range as equally probable.
4. The probability that m firms enter is then expressed as:

$$p(m) - 1/(Q + 1),\tag{8.3}$$

where Q = the range of the number of entries.

In this way the number of firms to enter is determined from a uniform distribution. Naturally, this approach is debatable, but it seems to be the best one, given only a few observations of entry. According to this method, entering firms are destined to the lowest class. Due to the small number of observations on entry, the possibilities of incorporating new features in the entry mechanism will be limited.

Validation of the Model

The model is represented in a computer program written in ALGOL, here documented in Appendix A. Due to limited funds, the model was run only for the sample SWED. It seemed more fruitful to run the model under different conditions for one sample than to run it once for each of the five samples. Three conditions were investigated:

1. The law of proportionate effect is assumed valid (P)
2. The law is not assumed valid, and firms can move into the fifth class (P_1^*)
3. The law is not assumed valid, and firms can leave the population from the fourth class (P_2^*) [7]

Table 8-1
Entry Probabilities Used for SWED

Probability that m firms enter	$\sim.167$
where m is	$2 \leqslant m \leqslant 7$
E (number of firms to enter)	4.5

The transition probabilities used here are those derived in chapter 5, whereas the entry probabilities were estimated by means of the short-cut method. The values of the latter are shown in table 8-1.

Each application was performed 100 times in a computer (IBM 360/75). The various runs differ only with respect to the random numbers used.

For each of the runs we computed:

1. The number of firms in class i
2. The total number of firms
3. The Pareto coefficient ρ [8]

The following summary measures of 100 runs for each application were derived:

1. The mean of the three types of variables obtained in every run
2. The standard deviation for the same values
3. The highest and lowest values of the same variables
4. The complementary distribution function, $F(S_i)$, corresponding to the mean in the classes

A sample of the output is shown in figure 8-3.

The model was first run in order to test the validity of the model. The following tests were undertaken:

1. Applications from 1956 to 1965, which is the period of estimation (Reproduction of historical conditions I)
2. Applications from scratch (zero firms) to the year in which the population has as many firms as in 1956 (Reproduction of historical conditions II)
3. Applications from 1965 to 1970 (Forecasting of future conditions)[9]

Of the two applications testing the reproduction of historical conditions, the first might cause objections since the testing was performed for the same period as the one used for estimation. This is a useful technical validation, however, as it tests whether or not the split into two mechanisms is appropriate.

The following discussion presents the mean and standard deviation for the

RESULT FROM THE SIMULATION RUNS

RUN	CLASS 1	2	3	4	5	6	7	8	9	10	TOT	RD
1	42	27	8	7	3	0	0	0	0	0	82	-1.15
2	38	24	12	7	3	0	0	0	0	0	84	-1.18
3	41	21	16	4	5	0	1	0	0	0	88	-1.16
4	44	28	13	4	1	1	0	0	0	0	91	-1.37
5	37	20	15	7	2	0	0	0	0	0	81	-1.30
6	46	21	13	3	2	0	0	0	0	0	85	-1.38
7	50	16	11	7	2	0	0	0	0	0	86	-1.29
8	36	19	10	8	2	0	0	0	0	0	79	-1.29
9	45	22	12	9	2	0	0	0	0	0	86	-1.26
10	39	23	12	4	1	1	0	0	0	0	79	-1.34
11	43	25	11	0	4	0	0	0	0	0	88	-1.19
12	44	29	11	5	2	0	0	0	0	0	86	-1.41
13	47	22	10	0	1	1	0	0	0	0	87	-1.49
14	40	24	10	5	1	1	0	0	0	0	89	-1.56
15	38	20	13	6	1	1	0	0	0	0	86	-1.24
16	47	28	14	5	4	0	1	0	0	0	99	-1.35
17	40	21	13	5	1	0	0	0	0	0	82	-1.53
18	46	20	15	4	1	0	0	0	0	0	98	-1.17
19	44	17	16	5	1	0	0	0	0	0	85	-1.10
20	34	21	20	4	1	0	0	0	0	0	88	-1.24
21	43	21	14	4	3	1	0	0	0	0	79	-1.61
22	49	25	13	3	2	1	0	0	0	0	92	-1.33
23	46	19	13	9	0	1	0	0	0	0	87	-1.25
24	52	13	14	5	1	0	0	0	0	0	79	-1.28
25	40	14	13	6	3	1	0	0	0	0	88	-1.41
26	42	26	11	5	3	0	0	0	0	0	87	-1.35
27	43	18	11	5	3	1	0	0	0	0	81	-1.25
28	44	22	11	7	1	0	0	0	0	0	83	-1.22
29	41	24	12	0	4	1	0	0	0	0	86	-1.75
30	37	23	16	4	3	0	0	0	0	0	82	-1.74
31	53	13	12	5	0	0	0	0	0	0	88	-1.49
32	46	26	14	3	1	1	0	0	0	0	82	-1.33
33	77	17	12	3	3	0	0	0	0	0	88	-1.21
34	51	24	15	4	4	0	0	0	0	0	88	-1.14
35	43	20	12	5	1	0	0	0	0	0	84	-1.21
36	42	23	11	3	2	0	0	0	0	0	89	-1.24
37	47	29	12	3	0	1	0	0	0	0	81	-1.42
38	35	23	13	6	1	1	0	0	0	0	84	-1.63
39	44	27	12	8	3	1	0	0	0	0	94	-1.27
40	41	26	11	4	2	1	0	0	0	0	87	-1.25
41	39	21	16	5	1	2	0	0	0	0	77	-1.14
42	53	19	12	4	4	1	0	0	0	0	92	-1.52
43	44	21	12	0	1	1	0	0	0	0	85	-1.26
44	42	25	12	7	4	0	0	0	0	0	81	-1.29
45	37	23	13	5	5	0	0	0	0	0	85	-1.32
46	33	22	13	5	2	1	0	0	0	0	77	-1.09
47	45	16	12	2	1	1	0	0	0	0	79	-1.26
48	45	21	13	6	4	1	0	0	0	0	85	-1.31
49	39	26	12	4	5	1	0	0	0	0	82	-1.05
50	41	18	7	6	2	2	0	0	0	0	78	-1.53
51	44	18	11	4	2	2	0	0	0	0	81	-1.16

55	39	28	14	3	3	0	0	0	0	0	86	-1.27
56	38	30	11	4	3	0	0	0	0	0	86	-1.25
57	42	23	11	5	2	1	0	0	0	0	84	-1.28
58	44	23	13	7	2	0	0	0	0	0	89	-1.33
59	46	19	17	5	1	1	0	0	0	0	89	-1.36
60	46	19	15	8	0	1	0	0	0	0	89	-1.43
61	40	33	9	2	2	0	0	0	0	0	86	-1.44
62	36	27	16	6	3	0	0	0	0	0	88	-1.23
63	47	20	13	8	1	1	0	0	0	0	90	-1.34
64	48	17	12	5	1	1	0	0	0	0	84	-1.31
65	49	22	10	8	0	0	0	0	0	0	89	-1.16
66	44	23	10	3	4	0	0	0	0	0	84	-1.13
67	37	18	15	4	3	0	1	0	0	0	78	-1.15
68	45	26	12	4	1	1	0	0	0	0	89	-1.35
69	49	22	13	3	2	0	0	0	0	0	89	-1.40
70	43	18	15	3	3	1	0	0	0	0	83	-1.24
71	41	23	10	7	3	1	0	0	0	0	85	-1.24
72	41	21	11	9	1	0	0	0	0	0	83	-1.48
73	44	17	11	4	5	0	0	0	0	0	81	-1.01
74	38	19	15	6	1	1	0	0	0	0	80	-1.32
75	37	25	12	4	3	0	0	0	0	0	81	-1.22
76	42	26	12	3	3	0	0	0	0	0	86	-1.26
77	45	17	16	5	2	1	0	0	0	0	86	-1.29
78	42	24	11	5	1	2	0	0	0	0	85	-1.14
79	37	24	12	5	3	0	0	0	0	0	81	-1.20
80	49	15	15	3	3	1	0	0	0	0	86	-1.24
81	41	21	13	8	2	0	0	0	0	0	85	-1.30
82	41	17	21	7	2	0	0	0	0	0	88	-1.33
83	36	24	14	6	3	0	0	0	0	0	83	-1.20
84	42	24	15	5	1	1	0	0	0	0	88	-1.36
85	38	21	16	5	1	0	0	0	0	0	81	-1.55
86	39	21	15	2	4	0	0	0	0	0	81	-1.15
87	46	20	18	3	2	0	0	0	0	0	89	-1.41
88	47	19	11	8	1	0	0	0	0	0	86	-1.50
89	40	22	14	3	1	1	0	0	0	0	81	-1.33
90	49	19	9	8	0	0	0	0	0	0	85	-1.13
91	40	18	15	3	3	0	0	0	0	0	79	-1.21
92	45	27	12	4	2	1	0	0	0	0	91	-1.31
93	39	17	15	3	3	0	0	0	0	0	77	-1.20
94	41	23	17	2	2	1	0	0	0	0	86	-1.31
95	39	23	13	4	3	1	0	0	0	0	83	-1.25
96	39	23	10	6	2	1	0	0	0	0	81	-1.26
97	38	27	8	8	2	0	0	0	0	0	83	-1.29
98	40	19	13	5	3	0	0	0	0	0	89	-1.21
99	42	30	3	8	4	0	0	0	0	0	88	-1.08
100	45	23	10	7	1	1	1	0	0	0	88	-1.11
MV	42.61	21.99	12.61	5.00	2.22	0.44	0.04	0.00	0.00	0.00	84.91	-1.28
SIGMA	4.35	3.87	2.69	2.05	1.19	0.55	0.20	0.00	0.00	0.00	3.67	0.13
%	100.00	49.82	23.92	9.07	3.18	0.57	0.05	0.00	0.00	0.00		
HIGH	52	33	21	9	5	2	1	0	0	0	94	-1.01
LOW	33	13	3	0	0	0	0	0	0	0	77	-1.63

Figure 8–3. An Example of Output

Table 8–2
Results Obtained in the First Application: 1956–65

Case	P	P_1^*	P_2^*
ρ (1965)			
Empirical value	1.34	1.34	1.51
Simulation: Mean	1.37	1.43	1.40
Standard deviation	.15	.17	.20
Lowest value	.96	.97	1.06
Highest value	1.61	2.08	1.98
Test value	– .34	–1.05	.49
Critical value (1% level of risk)	±5.84	±5.84	±9.93
N (1965)			
Empirical value	78	78	76
Simulation: Mean	75.62	75.63	74.30
Standard deviation	5.91	5.26	5.46
Lowest value	57	65	58
Highest value	86	87	87

Note: The difference between the empirical values and those given in table 4–4 is due to the fact that five firms were found to have been classified as merchandising firms during more than 50% of the period. The results did not seem to be influenced by this circumstance, however.

number of firms as well as the mean of the Pareto coefficient ρ. The latter estimate is used to test whether the empirical value may be regarded as a product of the model. The test can be performed using a t-test by formulating the variable:

$$t' = s_x(b_{yx} - \beta_{yx}) \quad \sqrt{N-2} / (s_y \sqrt{1-r^2}). \tag{8.5}$$

It is distributed according to Student's distribution with $N - 2$ degrees of freedom (see *Cramér,* 1958, p. 403). The value of this variable is then computed and compared with the critical value. As for N, we have no indication of the sampling distribution for this variable. Consequently, no test is performed for N. The results from the first application and test values are shown in table 8–2.

The table shows that the empirical value of ρ is very close to the values obtained from the simulation and that the tests do not yield significant results. In addition, it may be noted that the different cases (P, P_1^* and P_2^*) produce very similar results. In all three cases the empirical value is closer to unity than the value obtained from the simulation.

All mean values of N are lower than the empirical ones. With respect to classes, this particularly holds true for the lowest class. This may result from the use of the short-cut method for estimating entry probabilities.

Table 8-3
Results in the Second Application (Runs from Scratch)

Case		P	P_1^*	P_2^*
ρ (1956)				
Empirical value		1.78	1.78	1.78
Simulation:	Mean	1.94	1.98	1.99
	Standard deviation	.35	.34	.33
	Lowest value	1.21	1.31	1.35
	Highest value	2.76	2.76	2.76
Test value		- .74	- .93	- .97
Critical value (1% level of risk)		±9.93	±9.93	±9.93
Time Needed to Run from Scratch				
Simulation:	Mean	10.91	10.54	10.55
	Standard deviation	1.27	1.56	1.54
	Lowest value	8	7	7
	Highest value	15	14	14

The results obtained in the second application are shown in table 8-3. As in the first application, the empirical values of ρ are closer to unity than the value obtained from the simulation. Another similarity between this and the earlier application is that none of the test values is significant. In this application the number of firms in class 1 of the empirical distribution is below average, while the opposite holds true for the other classes. The values of ρ are very similar in the three cases, as is the time needed to run from scratch. As for this latter variable, it would have been interesting to see whether there really were no firms above S_m at the points of time indicated by the results. Unfortunately, however, statistics on the largest firms before 1956 are not readily available, which makes it difficult to perform this type of test. Instead, we have to stick to the test values of ρ, all of which are below the critical values, which means that there is no reason to reject the model.

Turning then to the last validation application, we obtained the results shown in table 8-4. The third application did not yield any significant results for ρ. Moreover, the mean values of N in the simulation are very close to the empirical values.

In conclusion, no reason to reject the model on the chosen level of risk has been encountered. As a consequence, we will now proceed to some forecasting runs.

Forecasting Runs

In forecasting we will first use the same estimates as in the validation runs, after which changes in the entry mechanism will be undertaken. The results of the first procedure are shown in table 8-5. In figure 8-4 below,

Table 8-4

Results Obtained in the Third Application: 1965-70

Case		P	P_1^*	P_2^*
$\rho(1970)$				
Empirical value		1.33	1.33	1.23
Simulation:	Mean	1.25	1.42	1.31
	Standard deviation	.11	.18	.15
	Lowest value	.98	1.03	1.06
	Highest value	1.52	1.66	1.76
Test value		.93	− .97	− .61
Critical value (1% level of risk)		± 4.60	± 4.60	± 9.93
$N(1970)$				
Empirical value		94	94	91
Simulation:	Mean	95.60	93.70	92.57
	Standard deviation	4.35	4.31	3.97
	Lowest value	85	84	83
	Highest value	106	103	102

Note: Unfortunately, *Ekonomen* discontinued its publication of statistics in 1967, the last figures published being those concerning 1966. However, publication of the data in question was taken up by *Veckans Affärer* in 1971. The distribution in the table is therefore taken from this magazine (August 5, 1971, pp. 30-39).

There were obvious differences between the samples of the two magazines. For this reason, the data in *Veckans Affärer* have been adjusted in order to correspond as closely as possible to *Ekonomen's* sample.

Table 8-5

Estimates of ρ and N in 1980 Compared with Earlier Values and Those in Steady State

Case		P	P_1^*	P_2^*
ρ				
Empirical values:	1956	1.78	1.78	1.78
	1965	1.34	1.34	1.51
	1970	1.33	1.33	1.23
Steady-state distribution		.80	−	.68
Simulation (1980):	Mean	1.07	1.31	1.01
	Standard deviation	.10	.04	.10
	Lowest value	.79	1.24	.81
	Highest value	1.33	1.43	1.42
N				
Empirical values:	1956	42	42	42
	1965	78	78	76
	1970	94	94	91
Simulation (1980):	Mean	125.99	127.97	118.68
	Standard deviation	7.53	5.47	5.85
	Lowest value	107	115	103
	Highest value	144	140	136

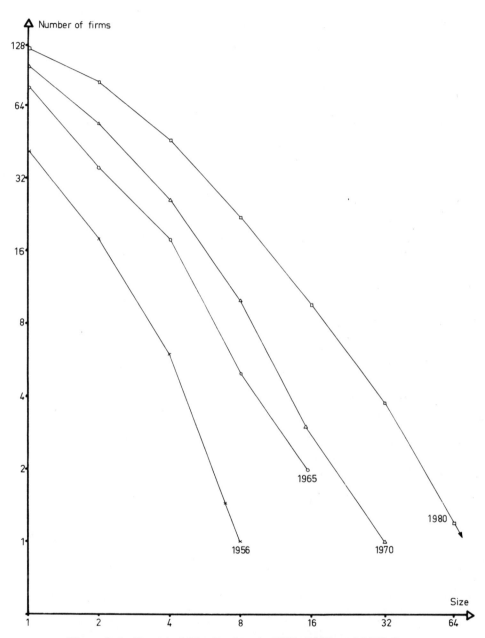

Figure 8–4. Empirical Distributions in 1956, 1965, and 1970 Compared with Averages from the Computer Simulation until 1980

the empirical distributions in 1956, 1965, and 1970 are compared with averages from the computer simulation until 1980. The latter distribution refers to the P-case.

Table 8–5 as well as figure 8–4 show an increasing degree of concentration—decreasing values of ρ—among the largest Swedish firms. The values of ρ approach unity and, consequently, also the values obtained for the steady-state distributions. The increase in concentration is most pronounced for P_2^* and P, whereas the change for P_1^* is fairly moderate. The simulation runs also indicate an increment of about twenty-five to thirty-five firms.

Of course, these results depend very much on the future development of probabilities of transition and entry. In order to check the sensitivity of our results, the model was also run using modified probabilities of entry. Two modifications were tried for the P-case: (1) the larger entry figures were given more weight, and (2) the smaller entry figures were given more weight. The entry probabilities were assumed to form a geometrical series having the quotients 2.0 and 0.5. Thus the probabilities shown in table 8–6 were used in the two alternative forecasting runs.

These two forecasting runs yielded the results shown in table 8–7. As for the values of ρ, i.e., the degree of concentration, the alternative forecasting runs did not produce results differing very much from the basic forecast. Thus these forecasts as well as the basic one indicate ρ-values approaching unity.

The differences with respect to the number of firms are more apparent. This is as was expected, since the alternative forecasting runs imply changes in the expected value of the number of firms to entry from 4.5 to 6.1 and 2.9, respectively.

The alternative forecasting runs indicate that the model is not extremely sensitive to the entry mechanism. Thus, if the number of entering firms does not change considerably, the basic forecast will probably suffice.

The alternative forecasting runs have not been performed for the cases where the law of proportionate effect is not valid. We have earlier noted the similarities between the cases P, P_1^* and P_2^* in the simulation runs. There is no reason to suspect that this is not also true when an alternative mechanism of entry is used.

Table 8–6
Entry Probabilities in the Alternative Forecasting Runs

Case	Size of entry	2	3	4	5	6	7	Expected value
1		1/63	2/63	4/63	8/63	16/63	32/63	6.1
2		32/63	16/63	8/63	4/63	2/63	1/63	2.9

Table 8–7
Estimates of ρ and N in 1980 in the Two Alternative Forecasting Runs

Case	Basic forecast	Alternative 1	Alternative 2
ρ			
Empirical values: 1956	1.78	1.78	1.78
1965	1.34	1.34	1.34
1970	1.33	1.33	1.33
Steady-state distribution	.80	.80	.80
Simulation: Mean	1.07	1.10	1.08
Standard deviation	.10	.10	.09
Lowest value	.79	.90	.90
Highest value	1.33	1.35	1.26
N			
Empirical values: 1956	42	42	42
1965	78	78	78
1970	94	94	94
Simulation: Mean	125.99	141.33	113.50
Standard deviation	7.54	5.26	4.75
Lowest value	107	126	102
Highest value	174	152	126

Conclusions

This chapter has introduced a simulation model using discrete size. The model has been tested for the sample of large Swedish firms with respect to its ability to reproduce historical conditions as well as to forecast future conditions. These tests have been performed both with and without the assumption of the law of proportionate effect. None of the tests have indicated any reason to reject the model.

In forecasting runs until 1980, the results indicate increases in concentration among the largest Swedish firms. Two alternative forecasting runs showed that the model is not very sensitive to the entry mechanism.

9

A Simulation Model Using Continuous Size

A model using discrete size uses only a rough measure of size, some information on size being left out. In some cases more complete information on firm sizes in the distribution may be desirable. In these instances a simulation model using continuous size can be used.

In this chapter we will discuss a model using continuous size.[1] The first section considers the model, whereas the following one discusses estimation procedures. In this chapter's third section, validation procedures are considered, and finally the fourth section presents results from empirical applications using the samples CAR and SHOE. Moreover, an example of the use of subjective probability estimates is shown.

The Processes of the Model

The models discussed in chapter 6 assume that the population is constant. As *Champernowne* (1956, p. 182) has pointed out, this assumption is often not realistic.[2] This deficiency can be remedied by using the simulation approach, which makes it possible to incorporate continuous size as well as exits and entries into the model.

The model in this chapter contains three basic components:

1. A mechanism determining changes in size
2. An exit mechanism
3. An entry mechanism

This means, in comparison with the model presented in the preceding chapter, that one mechanism is added: the exit mechanism. In the discrete model this mechanism is included in the transition process; here it is a process in itself.

In constructing the model, we will start with a simple version, making the same assumptions as in chapter 8, i.e., that (1) all firms are influenced by the same mechanism, and (2) the mechanisms are constant over time. These assumptions are important since they simplify the analysis to a large extent. However, they are not indispensable. Consequently, if there is any evidence that either or both of them ought to be discarded, this can be done.

The arguments in chapter 8 are also relevant here. Thus, if a firm has shown a high growth rate in earlier periods, it is possible to increase the probability of a high growth rate, and so forth.

The three parts of the continuous model are represented by probability distributions. The mechanism determining changes in size contains one distribution, whereas each of the other two mechanisms includes two probability distributions.

The mechanism determining changes in size is a distribution which expresses the probability of a certain relative change during one year. If the law of proportionate effect is assumed valid, as is done here in assumption (1) above, we may expect this distribution to have a normal shape (see the discussion in chapter 6). Support for this kind of distribution for growth rates has been found by *Hart & Prais* (1956, p. 170) and *Ansoff* (1959, p. 26).

The exit mechanism is similar to the entry mechanism in the previous chapter in that it contains one distribution determining the number of firms to exit and one distribution which determines their sizes.

With respect to the shapes of these distributions, there is no indication of a general shape for the first one, but the second may, according to studies of death risks of firms, be skewed to the left (see the section, "The Processes of the Model," in chapter 8). In all, the exit mechanism contains the following three steps:

1. Determine the number of firms (j) to exit during period t
2. Determine the "theoretical sizes" of firms to exit (S_{jt})
3. Extract, for each S_{jt}, the firm which is closest to S_{jt}

The entry mechanism also contains two probability distributions: (1) a distribution showing the probability of a certain number of firms entering, and (2) a distribution expressing the probability that an entering firm will be of a certain size.

That there is very little indication of the first distribution was emphasized in chapter 8. The second distribution, on the other hand, can be expected to belong to the skew family, implying that most entering firms are small.

The model can be summarized in a flow chart, as in figure 9-1.

Estimation Procedures

Introduction

Two approaches to estimating the probability distributions will be considered here. They imply the use of historical observations and subjective probability estimates, respectively.

The first of these methods has the advantage that a minimum of personal communication is needed. Thus, if data are available, the estimates desired are relatively easy to obtain. However, the use of such estimates presupposes that

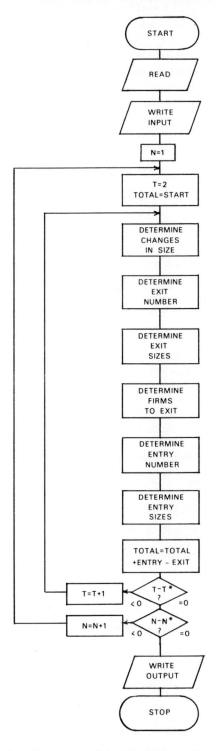

Figure 9-1. A Summary of the Model Using Continuous Size

the present trend will continue. From a planning point of view we are obviously more interested in cases where the trend changes. These cases may be caught if subjective probability estimates are used. The problems associated with such estimates include the choice of relevant experts and the interpretation of their probability estimates. It may even happen that the experts whom we wish to consult are unable to speak in probability terms. In all events, the validation of the model requires estimation of the distributions from historical data. Therefore, we will first discuss such estimates and then turn to subjective probability estimates.

Estimation from Historical Data

This section proposes estimation techniques when using historical data. Generally speaking, we suggest that observations should be equally weighted, although of course unequal weights may be assigned if a trend is observed.

Starting with the changes in size, we then suggest the following estimate:

$$p_g^* = \sum_{t=1}^{T} \sum_{i=1}^{w_t} a_{itg} / \sum_{t=1}^{T} w_t, \tag{9.1}$$

where

p_g = the probability of a change in size by $g\%$

a_{itg} = 1, if the largest integer value $< (G_{it} + 0.5) = g$, otherwise = 0

G_{it} = the change in size occurring in firm i during period t

w_t = the number of firms in period t

T = the period of estimation.

Turning to the first probability distribution incorporated in the exit mechanism, we would prefer to use the following estimate:

$$p(e) = (\sum_{t=1}^{T} b_t)/T, \tag{9.2}$$

where

$p(e)$ = the probability that e firms exit

b_t = 1, if e firms exit in period t, otherwise = 0.

This, however, might not be possible due to lack of data since every period yields only one observation on the number of firms that exit. One way to solve this problem is to assume all observed events to be equally probable and to use a uniform distribution. In that case the following alternative estimate is suggested:

$$p(e) = 1/(1 + J), \tag{9.3}$$

where
J = the range of the observed number of exits per year.

The probability that an exiting firm will have size S will be estimated as:

$$P^*(S) = \left(\sum_{t=1}^{T} c_t(S) \right) / \left(\sum_{t=1}^{T} d_t \right), \tag{9.4}$$

where
$P(S)$ = the probability that a firm which exits will have size S

$c_t(S)$ = 1, if a firm of size S exits in period t, otherwise = 0

d_t = the number of firms that exit in period t.

We meet the same problems concerning the entry mechanism as with the exit mechanism. Thus, we prefer:

$$p(m) = \left(\sum_{t=1}^{T} f_t \right) / T, \tag{9.5}$$

where
$p(m)$ = the probability that m firms enter

f_t = 1, if m firms enter during period t, otherwise = 0.

But we have to use:

$$p(m) = 1/(1 + Q), \tag{9.6}$$

where
Q = range of actual entries per year.

Finally, the probabilities that an entering firm will have size S can be estimated in the following way:

$$R^*(S) = \left(\sum_{t=1}^{T} h_t(S) \right) / \left(\sum_{t=1}^{T} q_t \right), \tag{9.7}$$

where

$R(S)$ = the probability that an entering firm will have size S

$h_t(S)$ = 1, if a firm of size S enters in period t, otherwise = 0

q_t = the number of firms entering in period t.

Subjective Probability Estimates

A relevant question when using the model for forecasting purposes is whether probability estimates from historical data really are appropriate. We face the same problem as that mentioned by Lundberg in a comparison between economic growth rates of different nations, when he warns against "more or less rash attempts to generalise our experiences of the growth processes of the 1950's and extrapolate these trends to cover the 1960's" (*Lundberg*, 1963, p. 113). Lundberg continues:

> the conditions for growth obtaining in the 1950's should be looked upon as unique, and we know that they are now undergoing a further change; I need hardly mention the new conditions arising from the creation of EEC and other trade blocks. *(ibid.)*

A suitable method for taking arguments such as these into account is the use of subjective probability estimates in forecasting. The task of obtaining such estimates involves several problems, the main one concerning what *Winkler & Murphy* (1968) refer to as normative and substantive standards of goodness.

The first of these concepts is related to the ability of a person to assess probability estimates, whereas the second has to do with his knowledge in the field considered. The problems stem from the fact that persons having high standards of goodness in both respects are very rare. Persons having a good knowledge of a certain industry might be unfamiliar with the probability concept, and vice versa.

Another problem connected with the use of subjective probability estimates is the judge's difficulty in reporting his "true" estimates. Such difficulties are likely to appear even with persons well acquainted with the probability concept. The main technique suggested to limit such difficulties involves the use of scoring rules (see, e.g., *Staël von Holstein*, 1970), but this technique has limitations as the events considered will occur in the distant future.

Another approach to help the experts is the use of specific questioning techniques. A number of these have been suggested (see, e.g., *Winkler*, 1967). That proposed most frequently involves sequential subdivisions of the outcome line and requires the judge to:

1. Divide the outcome line into two parts so that the actual outcome is equally likely to fall in either of them
2. Divide the right part into two parts as above in (1)
3. Divide the left part into two parts as above in (1)

Each of the four parts now obtained may then be divided into two parts, and so on. The first division will yield an estimate of the median of the probability distribution, whereas the second, third, and following divisions will yield information on the shape of the distribution.[3]

Using this technique, we may ask the following questions.[4]

Regarding changes in size:
> "What rate of change in size will be exceeded in P% of the coming F firm-years?"

Regarding exits:
1. "What number of exits/year will be exceeded in P% of the coming T years?"
2. "What size will P% of exiting firms exceed?"

Regarding entries:
1. "What number of entries/year will be exceeded in P% of the coming T years?"
2. "What size will P% of entering firms exceed?"

For all three mechanisms P is assumed to be 75, 50, and 25; but, of course, further subdivisions may be made if they are considered to improve the results.

When questioning, we must define in advance—preferably by written statements—the meaning of changes in size, exit, and entry. For example, it must be noted whether or not a merger is regarded as an exit for the acquired firm.

The questions can be posed with or without showing the experts the historical probability distributions. It might be a good procedure to start without showing these and then give the judges an opportunity to revise their estimates after having looked at them. A further step would be to run the model with subjective probability estimates and then discuss the results with the expert group.

The subjective probability estimates having been obtained from the various judges, we face the problem of aggregating the estimates. In this connection two approaches are discussed in *Winkler* (1968): the mathematical approach and the behavioral approach. The first approach involves the derivation of a weighted average of the different personal probability distributions, i.e., weighted averages of the parameters of the distributions are computed. The use of weights here means that those experts in whom we put more confidence are assigned more weight.

The second approach implies that feedback is given to the experts as to the expectations of others in order to get their estimates to converge. The Delphi technique can be a useful technique in this context.[5]

In the following empirical section we will mainly use estimates from historical observation. An example of the use of subjective probability estimates will, however, also be given.

Validation of the Model

The model is represented in a computer program written in *Fortran*, here documented in Appendix B.

In order to validate the model, it was applied to the two samples, CAR and SHOE. In so doing, the period 1952-61 was selected as the period for estimation. This period covers ten of the fifteen years for which we have collected data. The last five years are omitted in order to reserve data for testing the model.

The estimates of probabilities were derived according to the description in the subsection, "Estimation from Historical Data." These computations yielded the cumulative probability distributions with respect to changes in size shown in figure 9-2. The figure indicates that the distributions are approximately normal, a circumstance which was checked by plotting the observations on normal probability paper. Both distributions produce straight lines in this type of diagram. Tables 9-1 and 9-2 provide further illustrations of the two distributions.

Table 9-1 displays the growth rates having the highest probabilities, while table 9-2 shows the maximum likelihood estimates of the parameters of the distributions. Table 9-1 shows events with high probability to be in the range ± 5 percent for both populations. As for table 9-2, the contraction process in SHOE is emphasized by a lower mean and standard deviation as compared to CAR.

Concerning the number of firms to exit, the short-cut method of estimation was used, yielding the probability estimates shown in table 9-3. The probability distributions used to determine the sizes of exiting firms are displayed in figure 9-3.

As for the number of firms to enter, the short-cut method yielded the results shown in table 9-4.

Finally, figure 9-4 shows the probability distributions used to determine the sizes of entering firms.

The following information was gathered in each run in order to investigate the simulated distributions:

1. Mean of *ln*(sizes) – μ
2. Variance of *ln*(sizes) – σ^2
3. The number of firms – N

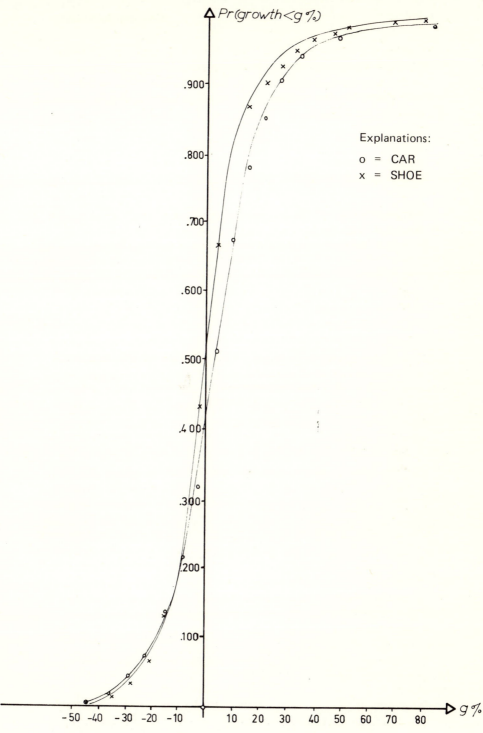

Figure 9–2. Probabilities of Size Changes in CAR and SHOE

119

Table 9–1
Growth Rates with the Highest Probabilities

Sample	CAR		SHOE	
Rank	Growth in %	Probability	Growth in %	Probability
First	0	.085	0	.116
Second	+4	.045	−4	.045
Third	−4	.037	+5	.033

Table 9–2
Estimates of the Parameters in the Distributions of Growth Rates (%)

Sample	CAR	SHOE
$\mu^*(G)$	5.154	1.290
$\sigma^*(G)$	23.042	19.037

Table 9–3
Probabilities that a Certain Number of Firms Leave the Two Populations

Sample	Pr(e firms exit)	where e is	E(e)
CAR	1/7 (~ .143)	$0 \leqslant e \leqslant 6$	3.0
SHOE	1/14 (~ .077)	$3 \leqslant e \leqslant 16$	9.5

The mean and standard deviation of these variables were computed in each sequence of runs. Tests were then performed to check whether the actual values could have been generated by the model. A t-test could be used with respect to μ since the sample distribution of the mean is normal. As for σ^2, we know that in a normal distribution $N(s^2/\sigma^2)$ is χ^2-distributed with $(N-1)$ degrees of freedom (see, e.g., Cramér, 1958, p. 382). Furthermore, for large values of n (degrees of freedom), we know that

$$\sqrt{2\chi^2} - \sqrt{2n-1}$$

has an approximately normal distribution with zero mean and unity variance (see Cramér, 1958, p. 250 ff.).[6] This gives us an opportunity to test the empirical values of variance for deviations from the estimated value in the simulation.

We have no indication of the sampling distribution for N. Consequently, no test can be performed for this variable. However, values of N will provide

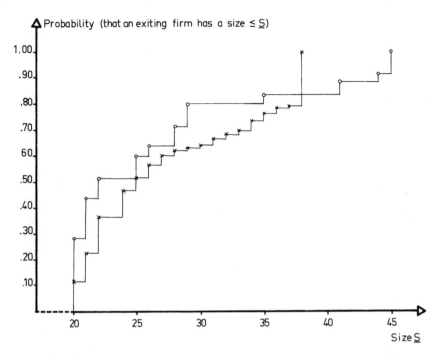

Figure 9–3. Probability Distributions Determining Sizes of Exiting Firms (x = CAR and o = SHOE)

Table 9–4
Probabilities that a Certain Number of Firms Enter the Two Populations

Sample	Pr(m firms enter)	where m is	E(m)
CAR	1/6 (~ .166)	$1 \leqslant m \leqslant 6$	3.5
SHOE	1/14 (~ .071)	$1 \leqslant m \leqslant 14$	7.5

information as to the face validity of the model. The same can be said concerning the final sizes in the last run, which are also included in the output.

The following validation runs were performed:

1. period 1952–61 (Reproduction of historical events I)
2. period 1961–66 (Reproduction of historical events II)
3. period 1961–69 (Forecasting of future events)[7]

This means that the same kinds of applications were made for this model as for the one in chapter 8. The number of runs here is also 100.

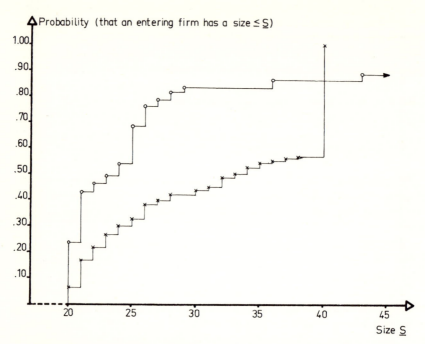

Figure 9-4. Probability Distributions Determining Sizes of Entering Firms (x = CAR and o = SHOE)

The applications produced output like that shown in figure 9-5, which refers to the second validation application for the sample SHOE.[8]

The results from the first application are shown in table 9-5.

The values of μ and σ are not significant on the chosen level of risk. Moreover, the actual values of N are fairly close to the mean in the simulation.

In the second application, the results shown in table 9-6 were obtained.

The conclusions as to μ and σ in the second application are similar to those in the first application. The differences between the empirical values for N and the mean in the simulation are larger, however.

Finally, table 9-7 shows the results of the third application.

Neither did the last validation application produce significant results for μ and σ, but larger differences for N between the simulated means and the empirical values were revealed.

In conclusion, we have found no reason to reject the hypothesis that the empirical values could have been generated by our model. It should be noted, however, that the empirical values of μ and σ are lower than the estimated mean in all cases. As for CAR, the empirical values of N are all higher than the mean estimates.

It must be pointed out that the fact that the model could not be rejected for the two samples used here does not guarantee similar results for any sample. Nevertheless, the two industries investigated here constitute two extremes with respect to growth (see "Data Sources" in chapter 2). For this reason the model might be applicable to a number of other industries.

The outcome of the validation applications implies that we are able to proceed, using the model for forecasting purposes.

Forecasting Runs

As in chapter 8, we will use the same estimates for forecasting as in the validation runs. Likewise, changes in the entry mechanism will be undertaken. Furthermore, an example of the use of subjective probability estimates will be shown.

When the estimates used earlier are employed, the results shown in table 9-8 are obtained for 1980. This table summarizes the mean of the different variables in the applications performed.

The applications indicate decreases in the mean values of μ and σ for CAR during the 1960s and increases during the 1970s. Thus the values of μ and σ should be expected to be somewhat larger in 1980 than in 1961. The simulation results also indicate increases in the number of establishments, a trend which will continue throughout the period.

Regarding SHOE, μ and σ are likely to be fairly constant during the 1960s as well as the 1970s. A slight decrease in the mean value of σ is indicated in 1980. The number of establishments producing shoes can be expected to decrease.

Estimates from Hart & Prais' method of σ are lower than the mean in the simulation for both samples.

Figure 9-6 displays the actual distributions in 1966 and one simulated distribution in 1980 for each sample (the last run). It illustrates the shifts in μ and σ. Moreover, the configuration of the plots indicates that lognormality is likely to persist.

In the alternative forecasting runs the entry probabilities have been changed in a way similar to that in chapter 8. The probabilities used in this context are shown in table 9-9. Both cases have been applied to both populations.

The applications using alternative entry probabilities yielded the results shown in table 9-10.

The first alternative for CAR yields reduced values of μ and σ relative to the basic forecast, whereas N is increased. The opposite obtains for the second alternative. In none of the cases does the degree of concentration change very much. Both alternative values of σ are larger than the estimate from Hart & Prais' method.

*** SIMULATION RESULTS ***

RUN	TOTAL	MY	SIGMA2
1	92	4.7446	0.5217
2	97	4.0260	0.6716
3	97	4.0081	0.7338
4	113	3.9441	0.6336
5	90	4.1258	0.6415
6	115	3.9120	0.5468
7	99	4.0640	0.6119
8	103	4.0844	0.6476
9	124	3.8586	0.5656
10	108	3.9123	0.6634
11	112	3.9268	0.4995
12	97	4.0762	0.6636
13	120	3.9378	0.6287
14	102	4.0091	0.5914
15	116	3.8841	0.5576
16	101	3.9883	0.6035
17	107	3.8778	0.6034
18	122	3.9191	0.6519
19	112	3.9709	0.7053
20	96	4.0383	0.6007
21	109	4.0275	0.5629
22	119	3.9326	0.6089
23	117	3.9013	0.6893
24	113	3.9413	0.5434
25	110	4.0051	0.5925
26	109	4.0299	0.7012
27	114	3.9419	0.6505
28	93	4.0180	0.5039
29	119	3.8587	0.5712
30	110	3.9975	0.5870
31	103	4.0896	0.5343
32	125	3.8658	0.5081
33	96	4.0345	0.7192
34	119	3.9738	0.6022
35	114	4.0102	0.6111
36	85	4.1804	0.6039
37	102	4.0087	0.6702
38	108	4.0058	0.6086
39	105	3.9726	0.5765
40	108	3.9966	0.5822
41	104	4.0545	0.5982
42	104	3.9851	0.6793
43	103	4.0017	0.5568
44	119	3.9442	0.5624
45	89	4.0546	0.7377
46	104	4.0150	0.6519
47	104	3.9706	0.5646
48	103	4.0156	0.5275
49	119	3.9348	0.6851
50	105	4.0694	0.4721
51	93	4.0319	0.7239
52	111	3.9460	0.5091
53	96	3.9916	0.5547
54	100	4.0275	0.5878
55	102	4.0030	0.5646
56	104	4.0395	0.4848
57	120	3.8827	0.5514
58	102	3.9960	0.5178
59	92	4.1304	0.5577

60	126	3.9142	0.5642
61	123	3.9378	0.5314
62	108	3.9533	0.6079
63	102	4.0598	0.6147
64	107	4.0044	0.6239
65	114	3.9174	0.5245
66	102	4.0592	0.5021
67	112	3.9991	0.5815
68	95	4.0444	0.6721
69	93	4.0782	0.7021
70	112	3.9536	0.6293
71	94	4.0727	0.6363
72	109	3.9568	0.6078
73	104	4.0149	0.5953
74	108	4.0360	0.5339
75	101	4.0631	0.7043
76	94	4.0326	0.7237
77	113	3.9657	0.6843
78	111	4.0099	0.5807
79	108	3.8953	0.6625
80	117	3.9718	0.5863
81	113	3.9405	0.5452
82	82	4.2318	0.5676
83	110	3.9955	0.5119
84	112	3.9123	0.5367
85	103	4.0043	0.5550
86	126	3.8730	0.5389
87	100	3.9632	0.7184
88	102	3.9858	0.7070
89	104	4.0062	0.6401
90	93	4.0805	0.5217
91	120	3.9441	0.5643
92	97	4.0491	0.6015
93	93	4.0624	0.6580
94	108	3.9683	0.6438
95	100	4.0288	0.5714
96	111	4.0091	0.6215
97	116	3.8643	0.5441
98	97	4.0332	0.5839
99	104	4.0362	0.6671
100	115	3.9441	0.6654

```
AVERAGE OF MY                   =    3.9956
STANDARD DEVIATION OF MY        =    0.0742
AVERAGE OF SIGMA2               =    0.6029
STANDARD DEVIATION OF SIGMA2    =    0.0642
AVERAGE OF NUMBER               =  106.3400
STANDARD DEVIATION OF NUMBER    =    9.7748
```

```
FINAL SIZES LAST RUN:
271  215  166   36   44   44  104   23  166  1101   35
 23  151  158   75   86   62   55   20   20    69   45
 28  165   29  103   46   99   67  171   55    41   93
 37   56   41  782  166   25  259  109   30    53   85
 36  400   41  105   98   20   49  109   48    70   59
 20   23   53  215   25   49  101  100   80   176  179
 63   35   27   58   24   44   33   20   20    88   83
 48   36   74   20   29  106   30  152   31    28   22
 20   27   30   54   48   22   44   30   35    27   37
 28   45   45   20   22   23   20   20   26    25   40
 23   40   40   34   32   50   25
```

```
TEST-VALUES: MY =-0.844 SIGMA =-2.060
```

Figure 9–5. An Example of Output

Table 9–5
Results from the First Validation Application: 1952–61

Sample	CAR	SHOE
μ		
Empirical value	4.495	3.931
Mean in simulation	4.881	4.077
Standard deviation in simulation	.322	.108
Test value	−1.199	−1.353
Critical value (1% level of risk)	±2.576	±2.576
σ		
Empirical value	1.461	.736
Mean in simulation	1.822	.788
Test value	−1.789	− .902
Critical value (1% level of risk)	±2.576	± 2.576
N		
Empirical value	49	115
Mean in simulation	40.590	113.080
Standard deviation in simulation	7.927	13.576

Table 9–6
Results from the Second Validation Application: 1961-66

Sample	CAR	SHOE
μ		
Empirical value	4.466	3.933
Mean in simulation	4.571	3.996
Standard deviation in simulation	.159	.074
Test value	− .660	− .844
Critical value (1% level of risk)	±2.576	± 2.576
σ		
Empirical value	1.340	.654
Mean in simulation	1.508	.777
Test value	−1.290	− 2.020
Critical value (1% level of risk)	±2.576	± 2.576
N		
Empirical value	80	91
Mean in simulation	50.780	106.340
Standard deviation in simulation	5.594	9.775

Regarding SHOE, alternative 1 implies an increase from the basic forecast in terms of μ and σ, whereas alternative 2 yields results similar to the basic forecast for both variables. For obvious reasons the mean value of number of firms decreases in both alternatives. Both mean values of σ are above that estimated by Hart & Prais' method, as was the case with the basic forecast.

Table 9-7
Results from the Third Validation Application: 1961-69

Sample	CAR	SHOE
μ		
Empirical value	4.341	3.883
Mean in simulation	4.651	4.057
Standard deviation in simulation	.175	.120
Test value	-1.771	- 1.450
Critical value (1% level of risk)	±2.576	± 2.576
σ		
Empirical value	1.332	.691
Mean in simulation	1.525	.790
Test value	-1.427	- 1.305
Critical value (1% level of risk)	±2.576	± 2.576
N		
Empirical value	75	66
Mean in simulation	50.960	100.730
Standard deviation in simulation	6.914	14.108

Note: There might be an overestimation of the empirical values of μ and σ and an underestimation of N. This originates from changes in the statistical classification in 1968. As for μ and σ, this bias is unfavorable to the model since larger empirical values yield larger test-values.

Table 9-8
Mean of the Different Variables at Four Points in Time

Variable	CAR	SHOE
μ		
1961	4.881	4.077
1966	4.571	3.966
1969	4.651	4.057
1980	4.896	4.290
σ		
1961	1.822	.788
1966	1.508	.777
1969	1.525	.790
1980	1.875	.758
Hart & Prais 1980	1.600	.590
N		
1961	40.590	113.080
1966	50.780	106.340
1969	50.960	100.730
1980	87.100	64.490

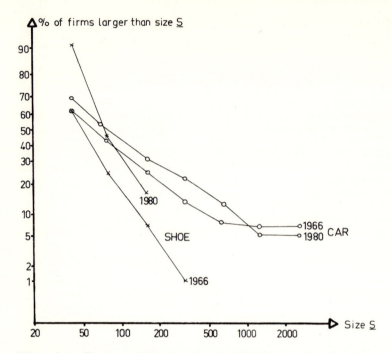

Figure 9-6. The Actual Distributions in 1966 Compared to Simulated Distributions in 1980 (Lognormal Probability Scale)

Table 9-9
Entry Probabilities Used in the Alternative Forecasting Runs

Case	1	2	3	4	5	6	Expected value
1	1/63	2/63	4/63	8/63	16/63	32/63	5.1
2	32/63	16/63	8/63	4/63	2/63	1/63	1.9

Up to this point we have run the model for historical data. We will now look at an example where subjective probability estimates are used.[9] This example illustrates how the model can be utilized to generate input to the strategic decision-making.

It is assumed that a corporation faces a decision as to which of two markets to enter. Both markets are quite unexplored and consist at the moment of two companies having sales of $0.4 million and $0.2 million, respectively. Information on earlier performance in the two markets is scarce, but available data do not indicate any significant differences up until now between the two markets.

The questions the corporation would like to have answered are:

Table 9–10
Results from Three Different Forecasting Runs

Variable	CAR	SHOE
μ		
Basic forecast	4.896	4.290
Alternative 1	4.651	4.455
Alternative 2	5.346	4.290
σ		
Basic forecast	1.875	.758
Alternative 1	1.783	.794
Alternative 2	1.938	.748
N		
Basic forecast	87.100	64.490
Alternative 1	106.060	46.000
Alternative 2	65.680	10.260

Table 9–11
Subjective Probability Estimates Used

Mechanism/Market	1	2
1. *Changes in Size (%)*		
$\mu(G)$	9	15
$\sigma(G)$	18	15
2. *Exit*		
$Pr(\text{Exit} = 0)$	.90	.75
$Pr(\text{Exit} = 1)$	.10	.25
$Pr(\text{Exit-size} > S)$	$.029\,S^{-2.38}$	$.096\,S^{-1.37}$
3. *Entry*		
$Pr(\text{Entry} = 1)$	.75	.25
$Pr(\text{Entry} = 2)$	.25	.75
$Pr(\text{Entry-size} > S)$	$.002\,S^{-3.84}$	$.029\,S^{-2.38}$

1. What will the structure within the two markets be five, ten, and twenty years from now?
2. What will be the expected size of a firm five, ten, and twenty years from now if it enters the market next year?
3. What are the chances of failure within the next five, ten, and twenty years for a firm entering the market next year? [10]

Subjective probability estimates have been collected by the corporation, and the aggregated estimates are those displayed in table 9–11.

In this application, besides μ, σ and N, the following variables were included in the output:

Table 9-12
Expected Values in the Runs Employing Subjective Probability Estimates

Variable	Period 5	Period 10	Period 20
μ			
Industry 1	-1.113	- .931	- .574
Industry 2	- .775	- .422	.220
σ			
Industry 1	.375	.505	.731
Industry 2	.466	.625	.973
Size of Entrant[a]			
Industry 1	.359	.546	1.328
Industry 2	.562	1.210	4.805
Market Size[a]			
Industry 1	2.831	6.190	19.448
Industry 2	5.213	14.418	69.025
Number of Firms			
Industry 1	7.870	13.470	24.890
Industry 2	9.380	16.800	31.830
Risk of Exit[b]			
Industry 1	.070	.090	.100
Industry 2	.210	.240	.240

[a]In $ million.
[b]This figure refers to the risk of exit before the period indicated by the columns.

1. Size of entrant
2. Market size
3. Risk of exit

The value of these variables was computed at three different points in time: in periods 5, 10, and 20.

The model was run 100 times with the results shown in table 9-12. The results shown in the table indicate that the values of all variables will be larger in market 2 on all three occasions. This means that larger sales are more likely in market 2, but also that the corporation will face competition from a larger number of firms. Moreover, market 2 is more risky with respect to the probability of failure. The decision will thus involve a tradeoff between sales prospects and the risks likely to be faced in the two markets.

Concluding Remarks

This chapter presented a continuous model under the law of proportionate effect which assumed a change in the number of firms. In comparisons with empirical distributions it has not been possible to reject the model. This circumstance might leave the impression that a search for ways of improving the model is superfluous. But, as was mentioned above, the results here do not guarantee similar results for other samples. As a consequence, before applying the model for purposes of forecasting, the same process of checking as the one used here must be carried out.

If the simple version of the model is rejected, we should look for ways of improving the model. The first aspect we should investigate, then, is the representation of growth probabilities. If the normal curve is not appropriate, we must find alternatives. The approach which immediately suggests itself is to handle large firms with one growth mechanism and small firms with another. This method does not always improve the applicability of the model with respect to a given population. In such cases the growth mechanism has to be revised still further. This stepwise procedure means that the model is gradually made more and more complex. This, in turn, implies that our preference for a simple over a complex model has to be modified.

The exit and entry mechanisms are other factors to be investigated, but in these cases it may be difficult to find other representations, as the number of observations on exit and entry that can be obtained is limited.

The possibilities of extending the model bring us to the general topic of future research in this area, which will be discussed in the next chapter. The final chapter also contains the general conclusions that can be drawn from this study.

Part 4
Conclusions

10 Conclusions and Suggestions for Future Research

The purpose of this chapter is to integrate the contents of earlier chapters. First, we will compare the different methods used in the study. We will then examine the inferences of the empirical distributions analyzed (USA-SWED, CAR, and SHOE). The final section contains some ideas about future research in this area.

The Methods

Various ways of measuring concentration were dealt with in chapter 2. The chapters which followed had to do with stochastic methods of analyzing changes in size distributions. Since the various methods were not discussed in direct relation to one another, it seems relevant to do so here.

Two theoretical distributions were investigated: the Pareto distribution and the lognormal distribution. They are the results of two stochastic models proposed by Simon & Bonini and Gibrat, respectively. The two models differ on some essential points, which are summarized in table 10-1.

The fact that Simon & Bonini's model deals with steady-state distributions implies that there is direct information on the future degree of concentration. This contrasts with the simple lognormal model, which only indicates that an increase will or will not occur. We may still wonder whether Simon & Bonini's model is the more advantageous, since it contains no information about the future location of the minimum size on the cost curve.

The relative size aspect seems to cause only slight inconvenience in using Simon & Bonini's model. For practical purposes, no significant difficulties are caused by this factor.

The assumption that firms enter the population is valid for most empirical distributions. Thus the treatment of entries in Simon & Bonini's model appears to be more relevant than that in the lognormal model.

The two models are not completely interchangeable, however, due to the difference in the valid size range. If we wish to study a complete size distribution, Simon & Bonini's model is not valid, since its use is limited to firms of extreme size. This point was also stressed in chapter 4, where we argued that the distribution must be truncated at a larger size than that proposed by Simon & Bonini. Nevertheless, the model is still appropriate for the firms that account for the

Table 10-1
Differences between the Models of Simon & Bonini and Gibrat

| | The model of | |
Question	Simon & Bonini	Gibrat
1. Discrete model?	Yes	No
2. Relative size?	Yes	Not necessarily
3. Steady state?	Yes	No
4. Entries?	Yes	No
5. Valid for a complete size range?	No	Yes
6. Resulting distribution?	Pareto	Lognormal

bulk of the total production. Due to such differences in validity, we used the Pareto distribution for the largest firms and the lognormal distribution for more complete samples. As a matter of fact, the question of entries did not render the lognormal distribution unfeasible. Some of the distributions under study showed a lognormal shape, even though there was a change in the number of firms. A counterbalancing factor in this connection might be that there are also deviations from the assumption concerning the law of proportionate effect. The two models have this premise in common. However, the validity of this law becomes more questionable when firms of small sizes are also taken into consideration.

Of course, the lognormal description—like the Pareto description—is an approximation. We might be able to find better descriptions, but we would expect them to be more complex than those discussed here, which are very easy to handle. This aspect constitutes one of the most important advantages of these two distributions. Hart expresses this idea with respect to the lognormal distribution as follows:

> The lognormal is only one of many theoretical distributions which may be applied to growth of firms; but because of its simplicity, and because of its many properties related to the familiar normal curve, it can serve as a useful tool of economic analysis until it is displaced by a superior form. (*Hart*, 1962, p. 35)

The fact that the distributions are based on a theory is also an advantage over other approaches to measuring concentration, as it provides us with better insight into the process of change. The theories also make it possible to develop new models, as has been done in this study.

These new models are mainly based on the use of Markov chains, a method discussed in chapter 5. The simulation models derived here offer a higher degree of flexibility than Markov chains. This is because it is usually easier to introduce the influence of a variety of important factors into simulation models. The limiting factor is the number of observations that can be used for estimation. The possibilities of differentiating between various groups of firms increase with this number.

A second advantage of the simulation method is that states other than steady states can be taken into account. This cannot be done when Markov chains are used in conjunction with entries.

Thirdly, the analysis can be performed beyond the present ranges of size. This cannot be accomplished with steady-state solutions of Markov chains unless the following two conditions are met: (1) the law of proportionate effect is valid, and (2) the probability of decline is larger than that of growth.

In the majority of the distributions investigated, the opposite relation is observed with respect to probabilities. In other words, the cases where steady states can be deduced with an unlimited number of classes are limited in number.

Because sizes larger than the present ones can seldom be included in the solutions, we are forced to disregard the very largest firms. It seems almost superfluous to point out that this is a serious drawback. One possible solution is the use of relative sizes, but then we run into the problem of interpretation discussed above. All these reasons appear to indicate that simulation possesses distinct advantages over the use of Markov chains.

We have proposed both a model using discrete size and one using continuous size. The discrete model was applied to the sample SWED. Simulated distributions were similar to empirical ones, and we found no reason to reject the model. The same could be said for the continuous model, which was applied to the samples CAR and SHOE. Thus both models appear to be feasible descriptions of the process generating skew size distributions. Should we wish to choose between the two models, the discrete model has simplicity in its favor. On the other hand, the continuous model might be more appropriate since it uses and yields more complete information on firm sizes. The choice between them can be made with reference to the specific information desired or to the circumstances in question. But nothing prevents us from applying the models alternately or together.

Before the models are used for the purpose of forecasting, validation has to be performed. Since validation runs for one set of samples do not guarantee validity for samples outside this set, the procedure has to be carried out for each new sample under consideration.

The discussion on methods may be summarized as follows:

1. In our study the distributions of large firms could be approximated by the Pareto distribution.
2. In the applications performed completer size distributions could be described in an acceptable way by the lognormal distribution.
3. Simulation may be expected to provide more information on future size

distributions than the derivation of steady-state solutions in Markov chains.
4.   Both the discrete and the continuous simulation models are consistent with
     empirical size distributions. We do not propose a choice between them,
     but emphasize the fact that validation must be performed before going on
     to forecasting runs.

The Empirical Distributions

The samples in this study functioned as a means of checking the relevance
of the different methods. This also provides information on concentration and
changes in size distributions.[1]

In the samples of large firms (USA-SWED), the highest degree of concentration is found for USA. Values close to the American ones are also obtained for
Swedish and Scandinavian firms. The results for USA* and EUR in 1965 indicate
that they have the lowest concentration of the five samples. It should be noted,
however, that in 1956 the level of concentration in these two areas was above
that in SCAN and SWED.

USA* and EUR have experienced a process of deconcentration. This can
be observed from the measures of mobility. Both Hymer & Pashigian's and
Adelman's indices show a high mobility in USA* and EUR. The opposite holds
true for the other three samples. It is hard to discern the reasons for the process
of deconcentration. One hypothesis with respect to USA* may be the effect
produced by entering non-European (especially Japanese) firms. As a matter of
fact, the populations of USA* and EUR coincide in 1956, but differ by twenty
firms in 1965. An increase in concentration might have been expected in EUR
due to a high incidence of mergers, which have also increased in number across
national boundaries (especially in the Common Market).[2]

A counterbalancing factor may be the establishment of American firms in
Europe—an hypothesis raised by *Servan-Schreiber* (1967), among others. If this
hypothesis is correct, these firms have competed heavily with the European
giants.

Concentration in 1965 is very similar in USA and SWED. This is consistent
with earlier studies involving more extensive samples than ours (see *Carling*, 1968,
and *Economic Concentration*, 1964–67). Despite differences in the size range
examined, the results emphasize the fact that the right-hand tail is the most
important part of a distribution.

As for the future, Gini coefficients are estimated to move toward values in
the range 0.35 to 0.45. This means that we can expect slight increases in concentration in the five areas.

The two Swedish establishment distributions contrast quite clearly with one
another with respect to concentration. The expanding sample (CAR) has a
higher degree of concentration in both of the years studied (1952 and 1966).

This fact is accentuated even more in 1966, since concentration decreased in SHOE while it increased in CAR.

The process of deconcentration in SHOE is manifested by the high value of the mobility index. There was also a rapid decline in the number of firms. The simple model of Gibrat is not consistent with a trend of decreasing concentration. It yields an increasing degree of concentration, as in the case of CAR. On the other hand, the modified model of Hart & Prais is compatible with increasing as well as decreasing concentration. This means that CAR is consistent with both models, while SHOE is only compatible with that of Hart & Prais. The main difference between the models is that Hart & Prais do not assume the law of proportionate effect to be valid. Thus tests performed did not indicate any reason to reject the law of proportionate effect for CAR, whereas doubts arose in the case of SHOE. Furthermore, the normal shape of the growth curve is less than pronounced for SHOE than for CAR, another indication of deviation from the law. The difference in growth rates is observed in terms of higher values of the mean and standard deviation for CAR. With respect to entries and exits, the average firm going into or going out of business is smaller in CAR than in SHOE. The differences are only slight, however.

Two methods were used to derive values of concentration for CAR and SHOE in 1980, i.e., Hart & Prais' method and the continuous simulation model. For CAR, the first method indicates a standard deviation of logarithms of 1.61, while the second yields a value of 1.88. This means that both methods predict an increase in concentration from the 1969 value, 1.33. For SHOE, the two methods produce values of 0.59 and 0.76. The empirical value in 1969, 0.65, is between these two values, which means that the two estimates differ in terms of the direction of change indicated.

Size distributions of Swedish establishments showed a lognormal shape with but one exception. Lognormality was also found for enterprises in socialist countries, which suggests that the processes under study here are not exclusively capitalistic phenomena.

Suggestions for Future Research

Concerning areas for future research related to the present subject, it seems reasonable first of all to devote resources to study the components of changes in concentration, i.e., changes in size as well as exits and entries of firms.

Factors which covary with changes in size should be sought via the collection of data and more profound examination of individual firms. This type of research can include case studies of different firms with regard to both internal and external growth. A similar approach may also be used to investigate the exits and entries of firms.

Studies of the following factors related to changes in concentration would also be worthwhile:

1.   Economic policy
2.   Other economic distributions
3.   Education

It is very important to get an idea of how economic policy affects concentration, particularly attempts to govern the level of concentration. Many actions in this context, while not aimed at influencing concentration, do so as a by-effect. As a matter of fact, it is often argued that economic policy seldom affects firms of different sizes in the same way. Liquidity squeezes, for instance, are considered to hit small firms harder than large ones. This is mainly because banks, for obvious reasons, are less apt to drop a big customer than a small one (see, e.g., *SOU*, 1967:6, p. 225). Another relevant aspect of this question is measures to be applied in order to limit concentration. Stimulating the establishment of new businesses is probably a very important method in this connection.

The second factor has to do with the relations between different economic size distributions, of which quite a few actually are skew. Important examples are the distributions of wealth and income (see, e.g., *Champernowne*, 1953) and the sales of different products in a company.[3] Studies have also revealed that the demand curves of single products over time are highly skewed (see, e.g., *Pessemier*, 1966, p. 202). The fact that skewness is exhibited in other economic distributions might lead us to suspect that relations exist between them. The link between distributions of products and firms, for instance, was analyzed in a study by *Wedervang* (1965). In turning to the relations between the different curves, however, we might experience some difficulty in explaining cause and effect. In fact, there can even be interplay between such distributions as those of firm sizes and wealth. Studies of relations such as these would probably produce some interesting results.

As for education, the third factor, *Denison* (1967) showed how important the general level of education is to the growth rates of countries. Similarly, we might suspect that the level of education in a firm is decisive for its growth. Questions such as the following might be looked into:

1.   Are firms with more highly educated executives more successful than others?
2.   Do firms with a young staff (= people with a more recent education) grow more rapidly?

The level of education might be very important, for example, in connection with the adoption of new techniques.

The arguments about education may be interpreted as a desire to regard firm distributions as a projection of the distribution of potential managers' talent. In an analysis of the stability of the income distribution, *Aitchison & Brown* (1957, p. 25) express a similar idea, suggesting that the distributions of attributes and talent are decisive.

All of the above research topics are general with respect to size distributions. They lead readily to the question of how they can be included in the simulation models derived in this study. The answer is that if relations of any kind are found, they can serve as a basis for dividing firms into groups experiencing different probabilities. Where simple models are not appropriate, this may be a very important development, facilitating the forecasting of future size distributions in various industries.

The areas for future research suggested are all related to factors influencing the processes studied here. These relations are not only of general interest, but can also be used in our models. Hopefully, the work reported on in this study will provide a basis for more extensive models in the future.

Appendixes

Appendix A: The Computer Program
Used for the Discrete Simulation Model

The program used for the discrete simulation model is written in ALGOL, consists of 227 cards and requires 11 K of storage space. A sample, consisting of the program used in the validation applications 1 and 3 as well as in the forecasting application, is presented on the following pages.

In the applications from scratch (validation application 2) the T-loop starting on the second page of the program is exchanged for a test concerning the number of firms. This test replaces the statement 'END' T-LOOP; on the third page of the program.

The program has been run on an IBM 360/75 and required between 27 seconds (second application, P_1^*) and 170 seconds (forecasting, P_1^*) of CPU-time to execute.

The program uses uniformly distributed random numbers. They were derived by means of the UNIFORM procedure.[1] In using this procedure, the values of the parameters are chosen according to the comments in the program. Table A-1 shows the values chosen.

Table A-1
Parameter Values Chosen

$D = 0$
$G = 999$
$Y = (N - 3) = 65\ 533$ the first time, otherwise $= 0$
$J = 253$
$N = 65\ 536$

The procedure is tested by the programmer with regard to two aspects: (1) distribution, and (2) independence. The results of these tests were consistent with the hypotheses that the numbers were uniformly distributed and that X_i and X_{i+j} are independent of the values of j.

The procedure originally contained 'OWN' 'REAL', but this is not allowed when using the IBM 360/75. Consequently, X is declared in the head of the main program.

```
'BEGIN'

'COMMENT'
****************************************************************
*    THIS IS A SIMULATION MODEL USING DISCRETE SIZE.    *
****************************************************************
END OF COMMENT;

'PROCEDURE' OUTLIST(A,B,C);
'VALUE' A;
'INTEGER' A;
'PROCEDURE' B,C;
'CODE';
'BEGIN'
'REAL' B,C,H,P,X;
'REAL' CLASS,LOG1,LOG2,LOG3,LOG4,RO,RUN;
'INTEGER' A,E,F,I,J,M,Q,R,S,T,U,V;
'INTEGER' START,TOT;
'REAL' 'ARRAY' FLOW,K(/1:6/);
'REAL' 'ARRAY' PAR(/1:10/);
'REAL' 'ARRAY' HIGH,LOW,MV,QUAD,SIGMA,SUM(/1:12/);
'REAL' 'ARRAY' IN,OUT,Z(/0:12/);

'PROCEDURE' UTPUTO(N,S); 'VALUE' N; 'STRING' S;
'INTEGER' N; 'CODE';
'PROCEDURE' UTPUT1(N,S,I1); 'VALUE' N,I1; 'STRING' S;
'INTEGER' N; 'REAL' I1; 'CODE';
'REAL' 'PROCEDURE' UNIFORM(D,G,Y,J,N);
'VALUE' D,G,Y,J,N;
'REAL' D,G;
'INTEGER' Y,N,J;

'BEGIN'
'IF' Y 'NOT EQUAL' O 'THEN' X:=Y/N;
X:=X*J;
X:=X-ENTIER(X);
UNIFORM:=X*(G-O)+D
'END' PROCEDURE UNIFORM;
H:=UNIFORM(OO,999,65533,253,65536);

SYSACT (1,8,50); SYSACT (1,12,1);
ININTEGER(O,V);
ININTEGER(O,S);
ININTEGER(O,START);
INREAL(O,B); INREAL(O,C);
INARRAY(O,Z);
INARRAY(O,FLOW); INARRAY(O,K);
```

```
OUTSTRING(1,'('RESULT FROM THE SIMULATION RUNS')');
SYSACT(1,14,2);
UTPUTO(1,'('4B'('CLASS')'')');

'FOR' A:=1 'STEP' 1 'UNTIL' 10 'DO'
'BEGIN'
UTPUT1(1,'('4B-ZD')',A);
'END';

UTPUTO(1,'('4B'('TOT')'6B'('RO')'')');
UTPUTO(1,'('/B'('RUN')'')');

'FOR' A:=1 'STEP' 1 'UNTIL' 12 'DO'
'BEGIN'
MV(/A/):=0;
SUM(/A/):=0;
QUAD(/A/):=0;
SIGMA(/A/):=0;
'END';

'FOR' R:=1 'STEP' 1 'UNTIL' S 'DO'
'BEGIN'
TOT:=START;

'FOR' A:=1 'STEP' 1 'UNTIL' 12 'DO'
'BEGIN'
OUT(/A/):=Z(/A/);
'END';

'FOR' T..=1 'STEP' 1 'UNTIL' V 'DO'
'BEGIN' F..=0.,

'FOR' A..=1 'STEP' 1 'UNTIL' 10 'DO'
'BEGIN' IN(/A/)..=OUT(/A/).,
OUT(/A/):=0;
'END'.,

OUT(/0/)..=0;
A..=1;
L:    M:=IN(/A/);

'FOR' I..=1 'STEP' 1 'UNTIL' M 'DO'
'BEGIN'
H:=UNIFORM(00,999,00,253,65536);
'IF' H<B 'THEN' OUT(/A-1/):=OUT(/A-1/)+1 'ELSE'
'IF' H>C 'THEN' OUT(/A+1/):=OUT(/A+1/)+1 'ELSE'
OUT(/A/):=OUT(/A/)+1;
'END';
```

```
F..=F+M.,
'IF' F 'LESS' TOT 'THEN'

'BEGIN' A..=A+1; 'GOTO' L 'END';

P:=UNIFORM(00,996,00,253,65536);
J:=1;
IL:    'IF' P>K(/J/) 'THEN'
'BEGIN'
J:=J+1; 'GOTO' IL;
'END';
E:=FLOW(/J/);
OUT(/1/):=OUT(/1/)+E;
TOT:=TOT+E-OUT(/0/);

'END' T-LOOP;

UTPUT1(1,'('/-ZZD')',R);
UTPUT1(1,'('9B-ZD')',OUT(/1/));
'FOR' A:=2 'STEP' 1 'UNTIL' 10 'DO'
'BEGIN'
UTPUT1(1,'('4B-ZD')',OUT(/A/));
'END';
UTPUT1(1,'('3B-ZZD')',TOT);

'BEGIN'
LOG1:=LOG2:=LOG3:=LOG4:=RO:=0;

'FOR' A:=1 'STEP' 1 'UNTIL' 10 'DO'
'BEGIN'
PAR(/A/):=0;
'END';

PAR(/1/):=TOT;
A:=2;
U:=0;
G: PAR(/A/):=PAR(/A-1/)-OUT(/A-1/);
U:=U+1;
'IF' PAR(/A/)>0 'THEN'

'BEGIN'
A:=A+1;
'GOTO' G 'END';
```

```
'FOR' A:=1 'STEP' 1 'UNTIL' U 'DO'
'BEGIN'
CLASS:=2**(A-1);
LOG1:=LOG1+LN(PAR(/A/))*LN(CLASS);
LOG2:=LOG2+LN(CLASS);
LOG3:=LOG3+LN(PAR(/A/));
LOG4:=LOG4+LN(CLASS)**2;
'END';

RO:=(U*LOG1-LOG2*LOG3)/(U*LOG4-LOG2**2);
UTPUT1(1,'('4B-ZD.DD')',RO);
'END' RO-PROCEDURE;

OUT(/11/):=TOT;
OUT(/12/):=RO;

'FOR' A:=1 'STEP' 1 'UNTIL' 12 'DO'
'BEGIN'
SUM(/A/):=SUM(/A/)+OUT(/A/);
QUAD(/A/):=QUAD(/A/)+OUT(/A/)**2;
'END';

'IF' R=1 'THEN'
'FOR' A:=1 'STEP' 1 'UNTIL' 12 'DO'
'BEGIN'
HIGH(/A/):=OUT(/A/);
LOW(/A/):=OUT(/A/);
'END';

'FOR' A:=1 'STEP' 1 'UNTIL' 12 'DO'
'BEGIN'
'IF' OUT(/A/)>HIGH(/A/) 'THEN' HIGH(/A/):=OUT(/A/);
'IF' OUT(/A/)<LOW(/A/) 'THEN' LOW(/A/):=OUT(/A/);
'END';

'END' R-LOOP;

RUN:=S;

'FOR' A:=1 'STEP' 1 'UNTIL' 12 'DO'
'BEGIN'
MV(/A/):=SUM(/A/)/RUN;
SIGMA(/A/):=SQRT(QUAD(/A/)/RUN-MV(/A/)**2);
'END';
```

```
UTPUTO(1,'('//'('MV')''')');

UTPUT1(1,'('11B-ZD.DD')',MV(/1/));
'FOR' A:=2 'STEP' 1 'UNTIL' 12 'DO'
'BEGIN'
UTPUT1(1,'('1B-ZD.DD')',MV(/A/));
'END';
UTPUTO(1,'('/'('SIGMA')''')');

UTPUT1(1,'('8B-ZD.DD')',SIGMA(/1/));
'FOR' A:=2 'STEP' 1 'UNTIL' 12 'DO'
'BEGIN'
UTPUT1(1,'('B-ZD.DD')',SIGMA(/A/));
'END';

PAR(/1/):=100;

'FOR' A:=2 'STEP' 1 'UNTIL' 10 'DO'
'BEGIN'
PAR(/A/):=PAR(/A-1/)-100*MV(/A-1/)/MV(/11/);
'END';

UTPUTO(1,'('/B'('%')'''');
UTPUT1(1,'('10B-ZZD.DD')',PAR(/1/));

'FOR' A:=2 'STEP' 1 'UNTIL' 10 'DO'
'BEGIN'
UTPUT1(1,'('-ZZD.DD')',PAR(/A/));
'END';

UTPUTO(1,'('/'('HIGH')''')');
UTPUT1(1,'('9B-ZD')',HIGH(/1/));

'FOR' A:=2 'STEP' 1 'UNTIL' 11 'DO'
'BEGIN'
UTPUT1(1,'('4B-ZD')',HIGH(/A/));
'END';

UTPUT1(1,'('4B-ZD.DD')',HIGH(/12/));
UTPUTO(1,'('/'('LOW')''')');
UTPUT1(1,'('10B-ZD')',LOW(/1/));

'FOR' A:=2 'STEP' 1 'UNTIL' 11 'DO'
'BEGIN'
UTPUT1(1,'('4B-ZD')',LOW(/A/));
'END';

UTPUT1(1,'('4B-ZD.DD')',LOW(/12/));

'END';
'END';
```

150

Appendix B: The Computer Program Used for the Continuous Simulation Model

This appendix presents the program used for the continuous simulation model. The program is written in *Fortran*, consists of 222 cards and requires 8 K of storage space. It is illustrated on the following pages.

Depending on the number of firms and the number of periods simulated, the program requires 21 seconds (CAR, Application 2) to 120 seconds (CAR, Application 1) of CPU-time to execute on an IBM 360/75.

The random numbers are generated by the subprogram RAND(Y). This procedure generates uniformly distributed random numbers.[1] In the mechanism for changes in size there is also a need for normally distributed random numbers. They are obtained by using the following procedure (see, e.g., *Martin*, 1968, p. 80):

$$W_i = \sum_{j=1}^{12} U_j - 6, \qquad (AB.1)$$

where

U_j = a random number between zero and unity.

(AB.1) produces random numbers, which are normally distributed with mean = 0 and variance = 1.

```
       FUNCTION RAND(Y)
       REAL*8 RN,Z,C/34359738368./,X/262147./
       REAL*8 Y
       Z=DABS(Y)
       IF(Z.EQ.0.0)Z=2.**29+3.
       Z=Z*X
       Z=DMOD(Z,C)
       RN=Z
       RN=RN/C
       Y=Z
       RAND=RN
       RETURN
       END
C
C
C      ****************************************************
C      * THIS MODEL SIMULATES FIRM GROWTH                 *
C      * THE MAIN COMPONENTS OF THE MODEL ARE:            *
C      * 1. A MECHANISM DETERMINING CHANGES IN SIZE.      *
C      * 2. AN EXIT MECHANISM.                            *
C      * 3. AN ENTRY MECHANISM.                           *
C      * THE MODEL IS CONTINUOUS WITH RESPECT TO SIZE.    *
C      ****************************************************
       REAL LOG1,LOG2,MY,MYSUM1,MYSUM2,MYMY
       REAL MYSIG,NORM,NUMBER
       REAL*8 Z
       INTEGER ENTRY,EXIT,RUN,S2,START
       INTEGER T,TOT1,TOT2,TOTAL,X1,X2
C
       DIMENSION U(12)
       INTEGER A1(18)
       DIMENSION B1(18)
       INTEGER A2(19)
       DIMENSION B2(19)
       INTEGER OUT(14)
       DIMENSION IN(12)
       DIMENSION PR1(14)
       DIMENSION PR2(12)
       INTEGER S1(200)
       INTEGER S(200)
C
       WRITE(6,900)
900    FORMAT(38X,' *** SIMULATION MODEL USING CONTINUOUS SIZE *** ')
       READ(5,1000)N,START,IT,GM,GS
1000   FORMAT(2I4,I3,2F7.4)
       READ(5,1050)PR1
1050   FORMAT(7F6.3)
       READ(5,1075)OUT,IN
1075   FORMAT(26I3)
       READ(5,1100)PR2
1100   FORMAT(12F6.3)
       READ(5,1125)(A1(I),B1(I),I=1,18)
1125   FORMAT(8(I4,F5.2))
       READ(5,1150)(A2(I),B2(I),I=1,19)
1150   FORMAT(8(I4,F5.2))
       WRITE(6,1200)N
1200   FORMAT(//40X,' NUMBER OF RUNS ',12X,' = ',I6)
       WRITE(6,1300)START
1300   FORMAT(40X,' NUMBER OF FIRMS IN PERIOD 1  = ',I5)
       WRITE(6,1400)IT
1400   FORMAT(40X,' NUMBER OF PERIODS SIMULATED  = ',I5)
       WRITE(6,1500)GM
1500   FORMAT(40X,' MEAN OF GROWTH ',13X,' = ',F10.4)
       WRITE(6,1600)GS
1600   FORMAT(40X,' STANDARD DEVIATION OF GROWTH = ',F10.4)
       WRITE(6,1700)(OUT(I),PR1(I),I=1,14)
```

```
1700    FORMAT(40X,' PR(< ',I2,' FIRMS EXIT) ',8X,' = ',F10.4)
        WRITE(6,1800)(IN(I),PR2(I),I=1,12)
1800    FORMAT(40X,' PR(< ',I2,' FIRMS ENTER) ',7X,' = ',F10.4)
        WRITE(6,1900)(A1(I),B1(I),I=1,18)
1900    FORMAT(40X,' PR(A FIRM THAT EXITS < ',I3,' ) =',3X,F8.4)
        WRITE(6,2000)(A2(I),B2(I),I=1,19)
2000    FORMAT(40X,' PR(A FIRM THAT ENTERS <',I3,' ) =',3X,F8.4)
        WRITE(6,2050)
2050    FORMAT(///,' START SIZES: ')
C
        Z=784631.
        READ(5,10)(S1(I),I=1,START)
10      FORMAT(48X,I5,27X)
        WRITE(6,12)(S1(I),I=1,START)
12      FORMAT(20X,11I8,3X)
        WRITE(6,14)
14      FORMAT(1H1,44X,' *** SIMULATION RESULTS ***')
        TOT1=0
        TOT2=0
        MYSUM1=0.0
        MYSUM2=0.0
        SIGS1=0.0
        SIGS2=0.0
        WRITE(6,2100)
2100    FORMAT(///,41X,' RUN ',3X,' TOTAL ',5X,' MY ',4X,' SIGMA2')
C
C       **************   START RUN-LOOP   **************
        DO 300 RUN=1,N
        TOTAL=START
        DO 15 I=1,START
15      S(I)=S1(I)
C       **************   START TIME-LOOP   **************
        DO 200 T=2,IT
C
C       *** START MECHANISM FOR CHANGES IN SIZE ***
        DO 25 I=1,TOTAL
        SI=FLOAT(S(I))
        NORM=0.0
        DO 24 J1=1,12
        U(J1)=RAND(Z)
24      NORM=NORM+U(J1)
        W=NORM-6.0
        G=(W*GS+GM)*0.01
        S(I)=INT((1.0+G)*SI)
        IF(S(I).LT.20)S(I)=20
25      CONTINUE
C       *** END MECHANISM FOR CHANGES IN SIZE ***
C
C       *** START EXIT MECHANISM ***
        J=1
        Y=(RAND(Z)+0.0000001)
30      IF(Y.LE.PR1(J))GO TO 50
40      J=J+1
        GO TO 30
50      EXIT=OUT(J)
        IF(EXIT.EQ.0)GO TO 70
        IF(EXIT.GT.TOTAL)EXIT=TOTAL
        DO 60 M=1,EXIT
        Y=(RAND(Z)+0.0000001)*0.999
        I=1
52      IF(Y.LE.B1(I))GO TO 55
        I=I+1
        GO TO 52
55      IM=A1(I)
        IS=0
```

```
               DO 58 I=1,TCTAL
               S2=S(I)
               IF(S2.EQ.0)GO TO 58
               IS=IS+1
56             X1=IABS(IM-S2)
               IF(IS.EQ.1)GO TO 57
               IF(X2.LE.X1)GO TO 58
57             K1=I
               X2=X1
58             CONTINUE
60             S(K1)=0
C              *** END EXIT MECHANISM ***
C
C              *** START ENTRY MECHANISM ***
7C             J=1
               Y=(RAND(Z)+0.0000001)
75             IF(Y.LE.PR2(J))GO TO 90
80             J=J+1
               GO TO 75
90             ENTRY=IN(J)
               IF(ENTRY.EQ.0)GO TO 100
               IX=TOTAL+1
               IZ=TCTAL+ENTRY
               DO 95 I=IX,IZ
               Y=(RAND(Z)+0.0000001)*0.999
               K=1
92             IF(Y.LE.B2(K))GO TO 93
               K=K+1
               GO TO 92
93             S(I)=A2(K)
95             CONTINUE
C              *** END ENTRY MECHANISM ***
C
100            TUTAL=IZ-EXIT
               DO 160 I=1,TOTAL
               IS=I
150            IF(IS.NE.IZ)GO TO 155
               IF(S(IZ).EQ.0)GO TO 160
155            IF(S(I).NE.0)GO TO 160
               S(I)=S(IS+1)
               S(IS+1)=0
               IS=IS+1
               GO TO 150
160            CONTINUE
200            CONTINUE
C              ***************  END TIME-LOOP  ***************
C
               LOG1=0.
               LOG2=0.
               DO 210 I=1,TOTAL
               S3=FLOAT(S(I))
               S33=ALOG(S3)
               LOG1=LOG1+S33
210            LOG2=LOG2+S33*S33
               TOT=1.0/FLOAT(TOTAL)
               MY=LOG1*TOT
               SIGMA=(TOT*LOG2-MY*MY)
               WRITE(6,250)RUN,TOTAL,MY,SIGMA
250            FORMAT(40X,I4,5X,I4,5X,F8.4,2X,F8.4)
               TOT1=TUT1+TOTAL
               TOT2=TOT2+TOTAL*TOTAL
               MYSUM1=MYSUM1+MY
               MYSUM2=MYSUM2+MY*MY
               SIGS1=SIGS1+SIGMA
               SIGS2=SIGS2+SIGMA*SIGMA
```

```
300      CONTINUE
C        *************** END RUN-LOOP ***************
C
C
         NUMBER=1.0/FLOAT(N)
         MYMY=MYSUM1*NUMBER
         MYSIG=SQRT(NUMBER*MYSUM2-MYMY*MYMY)
         SIGMY=SIGS1*NUMBER
         SIGSIG=SQRT(NUMBER*SIGS2-SIGMY*SIGMY)
         WRITE(6,450)MYMY
450      FORMAT(/40X,' AVERAGE OF MY ',17X,' =',F10.4)
         WRITE(6,500)MYSIG
500      FORMAT(40X,' STANDARD DEVIATION OF MY ',6X,' = ',F9.4)
         WRITE(6,550)SIGMY
550      FORMAT(40X,' AVERAGE OF SIGMA2',14X,' =',F10.4)
         WRITE(6,600)SIGSIG
600      FORMAT(40X,' STANDARD DEVIATION OF SIGMA2',3X,' = ',F9.4)
         SUMTOT=FLOAT(TOT1)*NUMBER
         WRITE(6,650)SUMTOT
650      FORMAT(40X,' AVERAGE OF NUMBER ',13X,' = ',F9.4)
         TOTS=SQRT(FLOAT(TOT2)*NUMBER-SUMTOT*SUMTOT)
         WRITE(6,700)TOTS
700      FORMAT(40X,' STANDARD DEVIATION OF NUMBER ',2X,' = ',F9.4)
         WRITE(6,750)
750      FORMAT(' FINAL SIZES LAST RUN: ')
         WRITE(6,800)(S(I),I=1,TOTAL)
800      FORMAT(20X,11I8,3X)
         STOP
         END
```

Notes

Chapter 1
Stochastic Models and Changes in Industrial
Structure

1. See, e.g., *Adelman* (1951), *Rosenbluth* (1955), and *Theil* (1967).
2. Simpler in relation to alternative approaches.
3. Note that the models are discrete or continuous with respect to *size*, and not *time*.

Chapter 2
Some Basic Concepts

1. The data from *Fortune* are also published in *Economic Concentration* (part 5, 1967).
   The data from *Ekonomen* can be found in the following issues: 1958:6, 18, 1959:20, 1960:17 and in the last issue (no. 20) of each of the years 1961–67.
2. This presentation does not deal with the firm in relation to firm behavior. For a discussion of this subject, see, e.g., *McGuire* (1964), chapter 2.
3. "Level of description" refers to the scale of size of the units examined; at a "low" level of description the focus is on smaller subunits (see *Ramström*, 1967, p. 391).
4. While *Ekonomen* includes data on companies without as well as companies with subsidiaries, *Fortune* covers only the largest firms *with* subsidiaries. Since consistency between samples was desired, only the data on companies with subsidiaries were used.
5. One possible deflator is the wholesaler's price index, used in chapter 5.
6. One exception in Sweden is the statistics on farm enterprises, as compensation for crop damage is based on statistics on registered farms.
7. For a discussion see, e.g., *Hotelling & Pabst* (1936) and *Friedman* (1937).
8. When applying the second rule to size distributions of firms, it is advisable to use logarithms of sizes.
9. See *Pareto*, 1896–97; *Lorenz*, 1905; *Gini*, 1911; *Ricci*, 1916; and *Dalton*, 1920.
10. Interesting papers in the 1950s on industrial concentration are, for example, *Adelman*, 1951; *Rosenbluth*, 1955; *Scitovsky*, 1955; and *Blair*, 1956. More recent discussions appear in *Theil*, 1967; *Hall & Tideman*, 1967; *Horowitz & Horowitz*, 1968; and *Hart*, 1971.
11. The firms are assumed to be ranked in descending order.
12. *Yntema* (1933, p. 428) presents a formula to compute the Gini coefficient which simplifies the calculation.
13. The index is usually called Herfindahl's index although Hirschman seems to

have been the first to derive it. Hirschman used the square root of the above expression to measure concentration of international trade (see *Hirschman*, 1945, p. 98 and Appendix A). Herfindahl, who presented his index in 1950 (see *Herfindahl*, 1950), derived it independently of Hirschman. *Rosenbluth* (1955, p. 60) seems to have been instrumental in popularizing the index. The index has also been mentioned as the Gini coefficient (see *Kindleberger*, 1962; *Michaely*, 1958; and *Tinbergen*, 1962). This brought Hirschman to comment on the paternity of the index, and he concluded: "The net result is that my index is named either after Gini, who did not invent it at all, or after Herfindahl, who reinvented it. Well, it's a cruel world." (*Hirschman*, 1964, p. 761).

14. The ranges for the measures are 0.008 and 0.010, respectively.

Chapter 3
Analytical Models in Review

1. Actually, it is frequently submitted that the probabilities of transition can be expected to vary with the short-run business cycle (see, for example, *Solow*, 1951; *Newman & Wolfe*, 1961; *Hart*, 1962; and *Mansfield*, 1962).
2. The Markov chain technique has been applied in various areas. Examples are brand-switching (*Robinson & Luck*, 1964), personnel planning (*Vroom & MacCrimmon*, 1968) and portfolio-selection (*Cyert & Thompson*, 1968). Standard references on this technique include *Feller* (1960), *Howard* (1960), *Kemeny & Snell* (1960) and *Takács* (1960).

Chapter 4
A Model Using Discrete Size

1. *Engwall* (1968) constitutes an earlier version of this chapter.
2. See *Simon* (1955), who refers to a proof by *Titchmarch* (1939, p. 58).
3. This test has to be used when the parameters of a Pareto distribution have been estimated.
4. Due to the skewness of the distributions considered, the actual mean value will be lower than (4.9). However, since this approximation is also used for entries and exits, we cannot say whether (4.9) will yield an overestimation or an underestimation of α.
5. Actually it does, but average costs are the same for sizes above S_m.
6. Standard Industrial Classification.
7. The mathematical expressions are given in *Quandt* (1966 a, p. 421).
8. Here the number of possible cases = number of samples x number of distributions x number of tests. In this case, it is $2 \times 3 \times 4 = 24$.
9. The values of k are 1.690 for USA, 1.413 for USA* and EUR, and 1.432 for SCAN and SWED.
10. It can be argued that the Kolmogorov & Smirnov test should be used

instead of a χ^2-test, but this test cannot be used here since the parameters are estimated. This is because critical values are not known (see *Massey*, 1951, p. 73).

In the χ^2-test the classes are determined so as to assign equal proportions to the different classes in the theoretical distribution.

The degrees of freedom are 7 since two parameters have been estimated.

11. The number of classes are reduced for USA*, EUR, and SWED in order to assign more than 10 observations to each class (cf. *Cramér*, 1945, p. 420).

12. Explanations: R_Y = Value of R from Yntema's formula

$R_{\rho 1}$ = Value of R from ρ and the actual value of M

$R_{\rho 2}$ = Value of R from ρ and $M = \rho$

13. 10 classes are used, which gives a critical value of 18.48 on a 1% level of risk.

14. 8 classes are used, which gives a critical value of 15.09 on a 1% level of risk with 5 degrees of freedom.

15. In a letter Wedervang has verified the fact that this figure refers to 1930. He stresses that his estimate is only a short-cut one. A regression analysis gives $\rho = 1.37$.

Chapter 5
The Transition Process

1. *Engwall* (1970 b) comprises an earlier version of this chapter.

2. See, for example, *Matthews* (1959, p. 205 ff.).

3. The state is usually observed once a year, which means that oscillating firms might be excluded.

4. Naturally, transitions between nonadjacent classes can be estimated in the same way as indicated by (5.1) and (5.3).

5. Of course, $p_{i \pm 2}$ must also be taken into consideration when they are greater than zero.

6. The same time unit as the one used to estimate the probabilities is applied. This is in most instances one year (see the discussion above in Chapter 2, "Period of Observation and Classification."

Returns to a class are not taken into consideration, as the average time spent in a class is computed.

7. Note that even a merger cannot bring about a growth of more than 100% due to the usual treatment of mergers (see "Empirical Section" below). Thus a merger between two firms may lead to an upward jump of two classes at the most.

8. Adelman's paper met criticism on the part of many economists, see *Blair* (1952), *Edwards et al.* (1952) and *Lintner & Butters* (1952). Replies were given in *Adelman* (1952 a and b).

9. This method is similar to one of the methods of testing the law of proportionate effect, i.e., plotting the logarithms of beginning and final sizes in a diagram. Prais also mentions the possibility of performing his test by using

absolute logarithms instead of absolute sizes (*Prais*, 1958, p. 269).

10. This question was also touched upon by *Blair* (1956, p. 355). He raised the question of whether "the selected number should consist of an *'identical'* or *'changing'* group."

11. Gort found that the geometric mean in most industries was close to 1. Horowitz pointed out that since the geometric mean obtained in empirical studies is only an estimate, it may be appropriate to introduce standard error into the analysis (*Horowitz*, 1964, p. 237).

12. Source: *Statistisk Årsbok* (1966, p. 477, table 489).

13. Of course, conclusions such as these depend very much on the method of classification and the time concept used. It is quite evident that chances for mobility increase with the number of classes and the length of the time period.

Chapter 6
Models Using Continuous Size

1. See *Kapteyn* (1903), *Kapteyn & Van Uven* (1916), and *Van Uven* (1917 a and 1917 b).

2. See, for example, *Aitchison & Brown*, (1957, p. 22 ff.).

3. This may be the result of a failure of available testing methods to discriminate between different distributions.

4. Actually, they also mention a fifth method. It is omitted here because it is a combination of the other four. See also *Tiku* (1968) for a discussion of estimation from censored samples.

5. Aitchison & Brown show that maximum efficiency in estimating μ is obtained for $d_j \sim 0.61$. The corresponding value for σ is 1.47.

6. See *Engwall* (1970 c, appendix 7).

7. This point has been stressed in numerous studies of the lognormal distribution; see, for example, *Aitchison & Brown* (1957, pp. 30 ff.) for a discussion and extensive references. The authors state that the question of testing is less compulsory when "some kind of generative process" is presumed.

8. Here the Kolmogorov & Smirnov test may be used since critical values for tests of normality with estimated mean and variance are given in *Lilliefors* (1967).

9. Symmetry in this instance is in relation to the diagonal that begins at the upper left-hand corner of the diagram.

10. For practical purposes we can use a table in *Aitchison & Brown* (1957, p. 154).

11. *Hart* (1957, p. 233) used a similar approach, but he did not apply any test statistics.

12. This simple expression is obtained in the following way:
    1. Determine S_n from the expression $z_n = (S_n - \mu)/\sigma$ or $S_n = \mu + \sigma z_n$
    2. Insert the parameters of the first moment distribution, then
       $$z^* = [(\mu + \sigma z_n - (\mu + \sigma^2)]/\sigma \text{ or } z^* = (z_n - \sigma).$$

13. See *Yntema* (1933, p. 428).
14. This is $1.34\sqrt{1.44}$, where 1.34 is the estimate of σ for 1966.
15. For a description see chapter 5, "Measures of Mobility."
16. The lower limit of the establishments in these figures is five workers. This means that somewhat more complete size distributions are taken into account than in previous applications. The source is *Statistisk Årsbok, 1968*, table 99, pp. 126–29.
17. Actually, the deviations appear to originate in the tobacco industry, since the beverage industry is approximately lognormal. This was checked and reported to me by Dr. Arne Gabrielsson of Uppsala University, who has conducted a study of the structure of the Swedish brewing industry (*Gabrielsson*, 1970).
    Concerning the discussion of cost curves, see the section, "The Assumptions," above.
18. As was mentioned above in the footnote to table 2–6, the lowest size considered is probably smaller in Poland than in the other three countries. However, since we here only are concerned with the shapes of the distributions, this difference is of no significance.
19. These ideas are further discussed in *Engwall* (1972 a).

Chapter 7
Simulation

1. This fact has been stressed by *Cohen & Cyert* (1961, p. 127). Advantages of the simulation technique other than those mentioned above can be found in, for instance, *Naylor et al.* (1966, p. 8).
2. See the Section, "Firm," in chapter 2 above for a definition of the concept "level of description."
3. See, for instance, *Naylor et al.* (1966, p. 10) for a discussion of this problem.
4. For a discussion of these two techniques of model-building—from simplicity toward complexity and vice versa—see *Ackoff* (1962, p. 118).

Chapter 8
A Simulation Model Using Discrete Size

1. See the section, "Estimating Transition Probabilities," in chapter 5 and *Adelman* (1958, p. 901).
2. *Engwall* (1970 a) constitutes an earlier version of this chapter.
3. See figure 3–1.
4. For a description see the preceding chapter.
5. It should be noted in this context that according to the studies of *Hart & Prais* (1956) it is difficult to find patterns in transition probabilities during the business cycle.

6. For example, it is much easier to ask an expert: "What will be the mean growth rate?" than "What is the chance that a firm from class 1 will move into class 2?"

7. Another case involving probabilities that changed with the business cycle was planned. But the period of estimation turned out to be too short for this application.

8. For a description of this measure, see chapter 4.

9. This application was performed in early 1969, when data for 1970 were not yet available.

Chapter 9
A Simulation Model Using Continuous Size

1. Like the model in chapter 8, the present model uses discrete time.

2. See the section, "The Assumptions," in chapter 6 for a discussion of this topic.

3. The second and third divisions actually yield estimates of the first and third quantiles of the distribution.

4. It could be argued that an easier way to obtain the information desired would be to ask the experts direct questions. However, in view of the difficulties concerning probability assessment mentioned above, this might not work out.

5. See, e.g., *Dalkey & Helmer* (1963) for a description.

6. Cramér considers values equal to or greater than 30 to be large values of *n*.

7. This forecast was performed in 1971, but before figures for 1969 were available for comparison with the simulation results. The choice of 1969 has to do with lags in statistics. The use of data from later years would have excluded the possibility of testing.

8. In addition to the results shown in the figure, the input values appear in the output.

9. This example has earlier been presented in *Engwall* (1971).

10. If the firm has activities in other markets, the information from (2) and (3) refers only to the activities in the market entered.

Chapter 10
Conclusions and Suggestions for Future Research

1. A presentation of the computer programs used can be found in *Engwall* (1972 b).

2. See *Näringsliv i omvandling* (1964).

3. This is often expressed as the 80%–20% rule, which means that 80% of the sales in a company stem from 20% of the products (see, e.g., *Magee*, 1967, p. 38).

Appendix A
The Computer Program Used for the Discrete
Simulation Model

1. See *Strome* (1967). This procedure was recommended to me by programmers at the Stockholm Computer Center.

Appendix B
The Computer Program Used for the Continuous
Simulation Model

1. This has been checked by Professor Terry C. Gleason of GSIA, Carnegie-Mellon University, Pittsburgh, Pennsylvania, who also provided me with the procedure.

List of Designations

The designations used in this study are listed below in alphabetical order. The page numbers at the right indicate the page where the designation is first used.

Designation	Meaning	Page
A	$\log (M/P)$	39
AR	a constant determined on the basis of the research activities in the population under study	98
$a_{i,i+1,t}$	the number of movements from class i to class $(i+1)$ during period t	50
a_{itg}	1, if the largest integer value $< (G_{it} + 0.5) = g$, otherwise $= 0$	114
$B(S, \rho + 1)$	the beta function of S and $(\rho + 1)$	31
b	estimate of the regression coefficient	46
b_t	1, if m firms enter during period t, otherwise $= 0$	99
C	a normalizing constant	31
C_i	the upper limit of class i	17
CR_n	the percentage of size units held by the n largest firms (concentration ratio)	18
$c_t(S)$	1, if a firm of size S exits in period t, otherwise $= 0$	115
D	the area to the left of the Lorenz curve	37
d_t	the number of firms that exit in period t	115
$E(g)$	the expected value of g	35
$E(G_i)$	the expected change in size in class i	35
$F(S)$	1 minus the distribution function	32
$f(S)$	the frequency distribution of firm sizes (S)	17
f_t	1, if m firms enter during period t, otherwise $= 0$	115
G	the net growth of all firms	32
G_{it}	the change in size occurring in firm i during period t	114
g	the net contribution of entering firms	32
$h_t(S)$	1, if a firm of size S enters in period t, otherwise $= 0$	116
H	entropy	20
HH	the index of Herfindahl	20
HR	relative measure of entropy	20
HT	the index of Hall & Tideman	20
I_t	the instability index proposed by Hymer & Pashigian in period t	57
J	the range of the observed number of exits per year	115
K	the total amount of size units	34
k	the ratio of the upper to the lower class limits in all classes	35

Designation	Meaning	Page
M	a constant	32
M_t	index of mobility proposed by Irma Adelman	55
m_{jt}	the proportion of units in state j at period t	51
N	the number of firms	20
N_u	the union of the number of firms at time t and $t-1$	57
n_i	the number of firms in class i	36
n_j	the entries into class j during period t	36
nc	the number of classes	35
P	the law of proportionate effect is valid	61
$[P]$	the transition matrix	27
P^*	the law of proportionate effect is not valid	61
$P(S)$	the probability that a firm that exits will have size S	115
$P_i = P_{it}$	the market share of the ith firm (at time t)	20
p_g	the probability of a growth by $g\%$	114
P_{ij}	the probability of going from class i to class j	26
p_j	the probability that one of the entering firms will go into the jth class	99
$p(e)$	the probability that e firms exit	114
$p(m)$	the probability that m firms enter	99
Q	the range of the number of entries	100
q_t	the number of firms entering in period t	116
R	the Gini coefficient	37
R_Y	value of Gini coefficient from Yntema's formula	43
$R_{\rho 1}$	value of Gini coefficient from ρ and actual value of M	43
$R_{\rho 2}$	value of Gini coefficient from ρ and $M = \rho$	43
R_σ	an estimate of the Gini coefficient from σ	71
RC	costs for research and development in % of turnover for the firm under study	98
$R(S)$	the probability that an entering firm will have size S	116
r	estimate of the correlation coefficient	77
r_n	n related to N where $n \leqslant N$	71
S	firm size	17
$S_i = S_{it} = S_t$	the size of firm i (at time t)	66
S_m	the size of the smallest firm in Simon & Bonini's model	32
S_n	the nth percentile of the size distribution	69
$\bar{S}_i$	the average size in class i	35
S_{jt}	the sizes of firms to exit	112
$\bar{S}_n$	the average size of the n largest firms	36
S_{max}	the size of the largest firm	34
s_b	standard deviation for the regression coefficient	46
s_x	standard deviation of x	104
s_y	standard deviation of y	104

Designation	Meaning	Page
$s.u.$	size unit	34
T	the length of the period observed	36
t	an index expressing time	26
t_i^a	the average time spent in class i in the actual distribution	55
t_i^m	the average time spent in class i in the industry with perfect mobility	55
u	a vector expressing the distribution in steady state	27
v_{jt}	a random variable, $E(v_{jt}) = 0$	51
v_t	a vector expressing the distribution in period t	27
v_{it}	the ith element of v_t	55
w_t	the number of firms in period t	114
$w(t-1)$	the joint growth potential of all firms in period $(t-1)$	93
$w_j(t-1)$	the growth potential of the jth firm in period $(t-1)$	93
z_n	the number of standard deviations in a distribution $N(0,1)$ which corresponds to r_n	71
α	g/G	32
$\alpha_{N/K}$	α estimated from N/K	47
β	the regression coefficient	44
ϵ_t	a random variable	66
μ	the arithmetic mean using logarithms	28
$\mu(G)$	the mean value of growth rates	120
μ_i^*	the estimate of μ by method i	69
ρ_{xy}	the correlation coefficient between $\log(S)$ in time t and $(t+1)$	67
ρ	a parameter in Simon & Bonini's model	31
σ	the standard deviation of logarithms of size	28
σ_ϵ^2	the residual variance	67
$\sigma(G)$	the standard deviation of growth rates	120
σ_i^a	the standard deviation for time	55
σ_i^*	the estimate of σ by method i	69

Bibliography

Ackoff, R. L., *Scientific Method: Optimizing Applied Research Decisions*. New York: Wiley, 1962.

Adelman, I. "A Stochastic Analysis of the Size Distribution of Firms." *Journal of the American Statistical Association* 53 (1958): December, pp. 893-904.

——, & Preston, L. E. "A Note on Changes in Industry Structure." *Review of Economics and Statistics* 42 (1960): February, pp. 105–08.

Adelman, I. "Long Cycles–A Simulation Experiment." In *Symposium on Simulation Models*, Hoggatt, A. C. and F. E. Balderston (eds.), Cincinnati: South-Western Publishing Co., 1963.

Adelman, M. A. "The Measurement of Industrial Concentration." *Review of Economics and Statistics* 33 (1951): November, pp. 269–96.

——. "Rejoinder." *Review of Economics and Statistics* 34 (1952): May, pp. 174–78 (1952 a).

——. "Rejoinder." *Review of Economics and Statistics* 34 (1952): November, pp. 356–64 (1952 b).

——. "Differential Rates and Changes in Concentration." *Review of Economics and Statistics* 41 (1959): February, pp. 68–69.

——. "Statement before the Subcommittee on Antitrust and Monopoly, September 10, 1964." *Economic Concentration* Pt. 1, Washington, D.C.: GPO, 1964, pp. 223–48.

——. "Comment on the 'H' Concentration Measure as a Numbers-Equivalent." *Review of Economics and Statistics* 51 (1969): February, pp. 99–101.

Aitchison, J., & Brown, J. A. C. "On Criteria for Descriptions of Income Distribution." *Metroeconomica* 6 (1954): December, pp. 88–107.

——. *The Lognormal Distribution with Special Reference to its Uses in Economics*. Cambridge: Cambridge University Press, 1957.

Allen, R. G. D. *Mathematical Analysis for Economists*. London: Macmillan, 10th ed. 1962 (1st ed. 1938).

Amstutz, A. E. *Computer Simulation of Competitive Market Response*. Cambridge, Mass.: MIT Press, 1970 (1st ed. 1967).

Ansoff, H. I. et al. "Planning for Diversification through Merger." *California Management Review* 1 (1959): Summer, pp. 24–35.

Ansoff, H. I. *Corporate Strategy*. New York: McGraw Hill, 1965.

Archer, S. H., & McGuire, J. "Firm Size and Probabilities of Growth." *Western Economic Journal*, 3 (1965): Summer, pp. 233–46.

Bain, J. S. *Barriers to New Competition*. Cambridge, Mass.: Harvard University Press, 1956.

——. *International Differences in Industrial Structure–Eight Nations in the 1950's*. New Haven & London: Yale University Press, 1966.

——. *Industrial Organization*. New York & London: Wiley, 2nd ed. 1968, (1st ed. 1959).

Balderston, F. E., & Hoggatt, A. C. *Simulation of Market Processes*. Berkeley, Calif.: Institute of Business and Economic Research, 1962.

Barba, V. "Die Konzentration der Industrieproduktion in Rumänien." *Management International Review* 9 (1969): 1, pp. 45–57.

Bates, J. "Alternative Measures of the Size of Firms." In *Studies in Profit, Business Saving and Investment in the United Kingdom*. P E. Hart, ed. Vol. I, London: Allen & Unwin, 1965, pp. 133–49.

Blair, J. M. "The Measurement of Industrial Concentration: A Reply." *Review of Economics and Statistics* 34 (1952): November, pp. 343–55.

——. "Statistical Measures of Concentration in Business: Problems of Compiling and Interpretation." *Bulletin of the Oxford University Institute of Statistics* 18 (1956): November, pp 351–72.

Bonini, C. P. *Simulation of Information and Decision Systems in the Firm.* Englewood Cliffs, N.J.: Prentice-Hall, 1963.

Carling, A. *Industrins struktur och konkurrensförhållanden*. Lund: Studentlitteratur, 1968 (see also SOU 1968:5).

Champernowne, D. G. "The Distribution of Income between Persons." Unpublished fellowship thesis deposited in King's College, Cambridge, 1937.

——. "A Model of Income Distribution." *Economic Journal* 63 (1953): June, pp. 318–51.

——. "Discussion on Paper by Mr. Hart and Dr. Prais." *Journal of the Royal Statistical Society* 119 (1956), ser. A, pt. 2, pp. 181-83.

——. *Uncertainty and Estimation in Economics*. Vol. 1–3. Edinburgh: Oliver & Boyd, 1969.

Charnes, A., & Cooper, W. W. *Management Models and Industrial Applications of Linear Programming*. Vol. 1. New York: Wiley, 1961.

Cohen, K. J. *Computer Models of the Shoe, Leather, Hide Sequence*. Englewood Cliffs, N.J.: Prentice-Hall, 1960.

——., & Cyert, R. M. "Computer Models in Dynamic Economics." *Quarterly Journal of Economics* 75 (1961): February, pp. 112–27. (Also reprinted in Cyert, R. M., & March, J. G. *A Behavioral Theory of the Firm*. Englewood Cliffs, N.J.: Prentice-Hall, 1963, pp. 312–25.

Collins, N. R., & Preston, L. E. "The Structure of Food-Processing Industries 1935–55." *Journal of Industrial Economics* 9 (1960–61): July, pp. 265–79.

——. "The Size Structure of the Largest Industrial Firms 1909–1958." *American Economic Review*, 51 (1961): December, pp. 986–1011.

Cramér, H. *Mathematical Methods of Statistics*. Princeton: Princeton University Press, 8th ed. 1958 (1st ed. 1945).

Cyert, R. M., & March, J. G. *A Behavioral Theory of the Firm*. Englewood Cliffs, N.J.: Prentice-Hall, 1963.

——, & Thompson, G. L. "Selecting a Portfolio of Credit Risks by Markov Chains." *Journal of Business* 41 (1968): January, pp. 39–46.

Dalkey, N., & Helmer, O. "An Experimental Application of the Delphi Method to the Use of Experts." *Management Science* 9 (1963): April, pp. 458–67.

Dalton, H. "The Measurement of the Inequality of Incomes." *Economic Journal* 30 (1920): September, pp. 348–61.

Davidson, S., et al. *An Income Approach to Accounting Theory*. Englewood Cliffs, N.J.: Prentice-Hall, 1964.

Denison, E. F. *Why Growth Rates Differ*. Washington, D.C.: Brookings Institution, 1967.

Duesenberry, J. S., et al. "A Simulation of the United States Economy in Recession." *Econometrica* 28 (1960): October, pp. 749–809.

Economic Concentration. Parts 1–5. Hearings before the Subcommittee on Antitrust and Monopoly of the Committee on the Judiciary United States Senate, 88th and 89th Session, Washington, D.C.: GPO, 1964–67.

Edwards, E., et al. "Four Comments on 'The Measurement of Industrial Con-Concentration'." *Review of Economics and Statistics* 34 (1952): May, pp. 156–74.

Einhorn, H. A. "Changes in Concentration of Domestic Manufacturing Establishment Output: 1939–1958." *Journal of the American Statistical Association* 57 (1962): December, pp. 797–803.

Ekonomen, Stockholm, 1947–.

Engwall, L. "Size-Distributions of Firms–A Stochastic Model." *Swedish Journal of Economics* 70 (1968): 3, pp. 138–59.

——. "A Simulation Model of Changes in Concentration." *Canadian Journal of Economics* 3 (1970): February, pp 39–61 (1970 a).

——. "Size-Changes Among Business Firms." *Metroeconomica* 22 (1970): Maggio-Agosto, pp. 133–48 (1970 b).

—— *Size Distributions of Firms*, Stockholm, 1970, dissertation (1970 c).

——. "Simulation of Market Changes." *Paper Presented at the Eighteenth International Meeting of the Institute of the Management Sciences*, March 1971.

——. "Concentration in Different Economic Systems." Research Report no. 67 from the Department of Business Administration. Stockholm University, rev. ed. February 1972 (1972 a).

——. "Six Computer Programs to be Used in Analysis of Industrial Concentration." Research Report no. 77 from the Department of Business Administration. Stockholm University, May 1972 (1972 b).

Feller, W. *An Introduction to Probability Theory and Its Applications*. New York: Wiley, 2nd ed. 1960 (1st ed. 1957).

Fisher, R. A. *Statistical Methods for Research Workers*. Edingburgh: Oliver, 6th ed. 1936 (1st ed. 1925).

Forrester, J. W. *Industrial Dynamics*. New York: MIT Press & Wiley, 1961.

Fortune. Chicago, Ill., 1893–.

Friedman, M. "The Use of Ranks to Avoid the Assumption of Normality Implicit in the Analysis of Variance." *Journal of the American Statistical Association* 32 (1937): September, pp. 675–701.

——. *Essays in Positive Economics*. Chicago: Chicago University Press, 1953.

Gabrielsson, A. "Koncentration och skalekonomi inom malt- och läskedrycksindustrin." Unpublished doctoral dissertation, Uppsala University, Uppsala, Sweden, 1970.

Galton, F. *Hereditary Genius*. London: MacMillan, (1892) reprint 1925.

Gibrat, R. "Une loi des répartitions économiques: l'effet proportionnel." *Bulletin de la statistique générale de la France* 19 (1930), p. 469 ff.

——. *Les inégalités économiques*. Paris: Sirey, 1931 (An English translation of pages 62–90 is available in *International Economic Papers* 7 (1957), pp. 53–70.)

Gini, C. *Vairabilità e mutabilità*. Cagliari: Dessi, 1911.

Gort, M. "Analysis of Stability and Change in Market Shares." *Journal of Political Economy* 71 (1963): February, pp. 51–63.

Hall, M., & Tideman, N. "Measures of Concentration." *Journal of the American Statistical Association* 62 (1967): March, pp. 162–68.

Hart, P. E., & Prais, S. J. "The Analysis of Business Concentration: A Statistical Approach." *Journal of the Royal Statistical Society* 119 (1956), ser. A, pt. 2, pp. 150–81.

Hart, P. E. "On Measuring Business Concentration." *Bulletin of the Oxford University Institute of Statistics* 19 (1957): August, pp. 225–48.

——, & Brown, E. H. P. "The Sizes of Trade Unions: A Study in the Laws of Aggregation." *Economic Journal* 67 (1957): March, pp. 1–15.

Hart, P. E. "The Size and Growth of Firms." *Economica, N.S.* 29 (1962): February, pp. 29–39.

——. "Entropy and Other Measures of Concentration." *Journal of the Royal Statistical Society* 134 (1971), ser. A, pt. 1, pp. 73–85.

Herfindahl, O. "Concentration in the Steel Industry." Unpublished Ph.D. dissertation, Columbia University, New York, N.Y., 1950.

Hirschman, A. O. *National Power and the Structure of Foreign Trade.* Berkeley, Calif.: University of California Bureau of Business and Economic Research, 1945.

——. "The Paternity of an Index." *American Economic Review* 54 (1964): September, pp. 761–62.

Hoggatt, A. C. "Simulation of the Firm." *IBM Research Paper.* RC-16, August, 1957.

Holt, C. C., et al. *Planning Production, Inventories and Work Force.* Englewood Cliffs, N.J.: Prentice-Hall, 1960.

Horowitz, A., & Horowitz, I. "Entropy, Markov Processes and Competition in the Brewing Industry." *Journal of Industrial Economics*, 16 (1968): July, pp. 196–211.

Horowitz, I. "A Note on the Hart-Prais Measures of Changes of Business Concentration." *Journal of the Royal Statistical Society* 127 (1964), pt. 2, pp. 234–37.

Hotelling, H., & Pabst, M. R. "Rank Correlation and Tests of Significance Involving No Assumption of Normality." *Annals of Mathematical Statistics* 7 (1936), pp. 24–43.

Howard, R. A. *Dynamic Programming and Markov Processes.* New York & London: Wiley, 1960.

Hymer, S., & Pashigian, P. "Turnover of Firms as a Measure of Market Behavior." *Review of Economics and Statistics* 44 (1962): February, pp. 82–87 (1962 a).

——. "Firm Size and Rate of Growth." *Journal of Political Economy* 70 (1962): December, pp. 556-69 (1962 b).

Ijiri, Y., & Simon, H. A. "Business Firm Growth and Size." *American Economic Review* 54 (1964): March, pp. 77–89.

——. "Effects of Mergers and Acquisitions on Business Firm Concentration." *Journal of Political Economy* 79 (1971): March/April, pp. 314-22.

Imai, K. "The Growth of Firms in Japanese Manufacturing Industries." *Hitotsubashi Journal of Commerce & Management* 4 (1966): November, pp. 59–77.

Iribarne, A. d'. "La population des établissements nouveaux." *Revue écono-mique* 18 (1967): November, pp. 975–1037.

Joskow, J. "Structural Indicia: Rank-Shift Analysis as a Supplement to Con-centration Ratios." *Review of Economics and Statistics* 42 (1960): February, pp. 113–16.

Kalecki, J. M. "On the Gibrat Distribution." *Econometrica* 13 (1945): January, pp. 161–70.

Kapteyn, J. C. *Skew Frequency Curves in Biology and Statistics.* Groningen: Nordhoof, 1903.

——, & Van Uven, M. J. *Skew Frequency Curves in Biology and Statistics.* Groningen: Hoitsema Bros., 1916.

Kemeny, J. G., & Snell, J. L. *Finite Markov Chains.* Princeton: Nostrand, 1960.

Kendall, M. G. "Discussion on Paper by Mr Hart and Dr. Prais." *Journal of the Royal Statistical Society* 119 (1956), ser. A, pt. 2, pp. 184–85.

Kindleberger, C. P. *Foreign Trade and the National Economy.* New Haven: Yale University Press, 1962.

Klein, L. *Economic Fluctuations in the United States, 1921–1941.* New York: Wiley, 1950.

Lilliefors, H. W., "On the Kolmogorov-Smirnov Test for Normality with Mean and Variance Unknown", *Journal of the American Statistical Association,* 62 (1967): June, pp. 399–402.

Lindbeck, A. *Svensk ekonomisk politik.* Stockholm: Aldus, 1968. See also: "Theories and Problems in Swedish Economic Policy in the Post-war Period." *American Economic Review* 58 (1968): June supplement, pp. 1–87.

Lindgren, B. W. *Statistical Theory.* New York: MacMillan, 3rd ed., 1962 (1st ed. 1960).

Lindgren, G. "Hur länge lever ett aktiebolag." *Balans* 1 (1949): 4, pp. 208–221.

Lintner, J., & Butters, J. K. "Further Rejoinder." *Review of Economics and Statistics* 34 (1952): November, pp. 364–67.

Lorenz, M. O. "Methods for Measuring Concentration of Wealth." *Journal of the American Statistical Association* 9 (1905): June, pp. 209–19.

Lundberg, E. "Economic Growth, Inflation and Stability–An International Comparison." *Skandinaviska Banken Quarterly Review* 44 (1963): 4, pp. 105–14.

Lydall, H. F. "The Growth of Manufacturing Firms." *Bulletin of the Oxford University Institute of Statistics* 21 (1959): May, pp. 85–111.

McAlister, D. "The Law of the Geometric Mean." *Proceedings of the Royal Statistical Society* 29 (1879), p. 367 ff.

McGuire, J. *Theories of Business Behavior.* Englewood Cliffs, N.J.: Prentice-Hall, 1964.

Magee, J. F. *Physical-Distribution Systems.* New York: McGraw Hill, 1967.

Mansfield, E. "Entry, Gibrat's Law, Innovation and the Growth of Firms." *American Economic Review* 52 (1962): December, pp. 1023–51.

Martin, F. F. *Computer Modeling and Simulation.* New York: Wiley, 1968.

Massey, F. J., Jr. "The Kolmogorov-Smirnov Test for Goodness of Fit." *Journal of the American Statistical Association* 46 (1951): March, pp. 68–78.

Matthews, R. C. O. *The Trade Cycle.* Cambridge: Nisbet, 1959.

Michaely, M. "Concentration of Exports and Imports: An International Comparison." *Economic Journal* 68 (1958): December, pp. 722–36.

Morand, J.–C. "Taille et croissance des entreprises." *Revue d'économie politique*, 77 (1967): 2, pp. 189–211.

Naylor, T. H., et al. *Computer Simulation Techniques*. New York: Wiley, 1966.

——, & Vernon, J. M. *Microeconomics and Decision Models of the Firm*. New York: Harcourt, Brace & World, 1969.

Newman, P., & Wolfe, J. N. "A Model for the Long-run Theory of Value." *Review of Economic Studies* 29 (1961): October, pp. 51–61.

Niehans, J. "Eine Messziffer für Betriebsgrössen." *Zeitschrift für die Gesamte Staatswissenschaft* 111 (1955): 3, pp. 529–42. (An English translation is available in *International Economic Papers* 8 (1958), pp. 122–32.)

Näringsliv i omvandling. Stockholm: SNS, 1964.

Pareto, V. *Cours d'économie politique*. pts. 1–2. Lausanne & Paris: Rouge & Pichon, 1896–97.

Pessemier, E. A. *New Product Decision: An Analytical Approach*. New York: McGraw Hill, 1966.

Popper, K. R. *The Logic of Scientific Discovery*. London: Hutchinson, 1959 (a translation of *Logik der Forschung*. Vienna: Springer, 1935).

Prais, S. J. "Measuring Social Mobility." *Journal of the Royal Statistical Society* 118 (1955), ser. A, pp. 56–66.

——. "The Statistical Conditions for a Change in Business Concentration." *Review of Economics and Statistics* 40 (1958): August, pp. 268–72.

Preston, L. E., & Bell, E. J. "The Statistical Analysis of Industry Structure: An Application to Food Industries." *Journal of the American Statistical Association* 56 (1961): December, pp. 925–32.

——, & Collins, N. R. *Studies in a Simulated Market*. Berkeley, Calif.: Institute of Business and Economic Research, 1966.

Quandt, R. E. "On the Size Distribution of Firms." *American Economic Review* 56 (1966): June, pp. 416–32 (1966 a).

——. "Old and New Methods of Estimation and the Pareto Distribution." *Metrika* 10 (1966): 1, pp. 55–82 (1966 b).

——. "Lognormality and Concentration: A Comment." *American Economic Review* 58 (1968): December, pp. 1367–70.

Ramström, D. *The Efficiency of Control Strategies*. Uppsala: Almqvist & Wiksell, 1967.

Ricci, U. *L'indice di variabilità e la curve dei redditi*. Rome: Athenaeum, 1916.

Richman, B. M. *Soviet Management*. Englewood Cliffs, N.J.: Prentice-Hall, 1965.

Robichek, A. A., & Myers, S. C. *Optimal Financing Decisions*. Englewood Cliffs, N.J.: Prentice-Hall, 1965.

Robinson, P. J., & Luck, D. J. *Promotional Decision Making. A Study in Marketing Management*. New York: McGraw Hill, 1964.

Rosenbluth, G. "Industrial Concentration in Canada and the United States." *Canadian Journal of Economics and Political Science* 24 (1954): August, pp. 332–46.

——. "Measures of Concentration." *Business Concentration and Price Policy*.

National Bureau Committee for Economic Research, Princeton: Princeton
University Press, 1955, pp. 57–95.

Samuels, J. M. "Size and Growth of Firms." *Review of Economic Studies*
32 (1965): April, pp. 105–12.

Scherer, F. M. *Industrial Market Structure and Economic Performance*. Chicago,
Ill.: Rand McNally, 1970.

Scitovsky, T. "Economic Theory and the Measurement of Concentration."
Business Concentration and Price Policy. National Bureau Committee for
Economic Research, Princeton: Princeton University Press, 1955, pp.
100–13.

Servan-Schreiber, J.-J. *Le défi américain*. Paris: Denoël, 1967.

Shannon, C. E., & Weaver, W. *The Mathematical Theory of Communications*.
Urbana, Ill.: University of Illinois Press, 1949.

Shubik, M. "Simulation of the Industry and the Firm." *American Economic
Review* 50 (1960): December, pp. 908–19.

Silberman, I. H. "Application of the Lognormal Distribution to Industrial
Concentration." Unpublished Ph.D. dissertation, Massachusetts Institute of
Technology, Cambridge, Mass., 1964.

——. "On Lognormality as a Summary Measure of Concentration." *American
Economic Review* 57 (1967): September, pp. 807–31.

——. "Lognormality and Concentration: Reply." *American Economic Review*
58 (1968): December, pp. 1370–71.

Simon, H. A. "On a Class of Skew Distribution Functions." *Biometrica* 42
(1955): December, pp. 425–40. (Also reprinted in Simon, H. A. *Models of
Man*. New York: Wiley, 1957, pp. 145–64.)

——, & Bonini, C. P. "The Size Distribution of Business Firms." *American
Economic Review* 48 (1958): September, pp. 607–17.

Simon, H. A. "Comment: Firm Size and Rate of Growth." *Journal of Political
Economy* 72 (1964): February, pp. 81–2.

Singh, A., & Whittington, G. *Growth, Profitability and Valuation*. London:
Cambridge University Press, 1968.

Solow, R. "Some Long-run Aspects of the Distribution of Wage Incomes,
Abstract." *Econometrica* 19 (1951): 3, pp. 333–34.

SOU 1967:6. *Finansiella långtidsperspektiv*. Stockholm, 1967.

SOU 1968:5. *Industrins struktur och konkurrensförhållanden*. Stockholm, 1968.
(See also Carling, 1968.)

Staël von Holstein, C.-A. S. *Assessment and Evaluation of Subjective Probability
Distributions*. Stockholm: Norstedts, 1970.

Statistisk Årsbok (Statistical Yearbook of Sweden), Stockholm, 1914–.

Statistickà ročenka 1968. Praha: SNTL·ALFA, 1968.

Steindl, J. *Random Processes and the Growth of Firms*. London: Griffin, 1965.

Strome, W. W. "Algorithm 294." *Communications of the ACM* 10 (1967): 1,
p. 40.

Svenska aktiebolag (Directory of Swedish Incorporated Companies). Stockholm:
Norstedts, 1902–.

Sveriges industri. Stockholm: Esselte, 1967.

Takács, L. *Stochastic Processes*. London & New York: Methuen & Wiley, 1960.

Telser, L. G. "Least-Square Estimates of Transition Probabilities." *Measurement in Economics*. Stanford: Stanford University Press, 1963, pp. 270–92.

Theil, H., & Rey, G. "A Quadratic Programming Approach to the Estimation of Transition Probabilities." *Management Science* 12 (1966): May, pp. 714–21.

——. *Economics and Information Theory*. Amsterdam: North Holland Publishing Co., 1967.

Tiku, M. L. "Estimating the Parameters of Log-normal Distribution from Censored Samples." *Journal of the American Statistical Association* 63 (1968): March, pp. 134–40.

Tinbergen, J. *Shaping the World Economy*. New York: Twentieth Century Fund, 1962.

Titchmarch, E. C. *The Theory of Functions*. Oxford: Oxford University Press, 2nd ed. 1939 (1st ed. 1932).

Van Uven, M. J. "Logarithmic Frequency Distributions." *Proceedings from the Academy of Sciences Amsterdam* 19 (1917), p. 533 ff. (1917 a).

——. "Skew Frequency Curves." *Proceedings from the Academy of Sciences Amsterdam* 19 (1917), p. 670 ff. (1917 b).

Veckans Affärer 6 (1971): August 5, pp. 30–39.

Velkovíc, L. *Le développement économique de la Yugoslavie*. Beograd, 1968.

Vroom, V. H., & MacCrimmon, K. R. "Toward a Stochastic Model of Managerial Careers." *Administrative Science Quarterly* 13 (1968): June, pp. 26–46.

Wallis, W. A., & Roberts, H. V. *Statistics—A New Approach*. New York: The Free Press of Glencoe, 1956.

Wedervang, F. *Development of a Population of Industrial Firms*. Bergen: Universitetsforlaget, 1965.

Winkler, R. L. "The Assessment of Prior Distributions in Bayesian Analysis." *Journal of the American Statistical Association* 62 (1967): September, pp. 776–800.

——. "The Consensus of Subjective Probability Distributions." *Management Science* 15 (1968): October, pp. B61–B75.

——, & Murphy, A. H. " 'Good' Probability Assessors." *Journal of Applied Meteorology* 7 (1968), pp. 751–58.

Woroniak, A. "Industrial Concentration in Eastern Europe: The Search for Optimum Size and Efficiency." *Notwendigkeit und Gefahr der wirtschaftlichen Konzentration*. Basel: Kyklos, 1969, pp. 265–84.

Yntema, D. B. "Measures of Inequality in the Personal Distribution of Wealth or Income." *Journal of the American Statistical Association* 28 (1933): December, pp. 423–33.

Author Index

Subject Index

About the Author

Lars Engwall received his Ph.D. at Stockholm University, Sweden. He is presently an Associate Professor (Universitetslektor) of Business Administration at Stockholm University. His research has included studies of industrial structure, firm growth, corporate boards, and strategic planning.